Braced Oil Dampers for Buildings

The high damping force of oil damper vibration control systems in buildings offers relative safety during earthquakes and at a low environmental cost. The oil damper is connected using braces in series and is installed with pins to the building structure, which minimizes the impact on the building structure and allows installation and removal with ease.

This is the first book detailing experimental work and cases of the use of oil dampers in buildings. It shows their effectiveness by detailing the testing and analysis of buildings with them placed at braced positions, exposed to small amplitude vibration, and without them.

- The only comprehensive presentation of oil damper vibration control
- Covers analysis and design, with case studies and details of testing and experimental results

The book is organized systematically to suit students and junior professionals in structural design as well as more specialist engineers. An additional Python code sample is available online for learning the behaviors of oil dampers (Jupyter Notebook environment required).

Osamu Takahashi is a Professor at Tokyo University of Science, founder of Science Kozo Inc., and an advisor to the New International Structural Design & Engineering Challenge Association, in Japan.

Atsuki Yokoyama works in the Research and Development Department of Sanwa Tekki Corporation and has a PhD in Engineering from Tokyo University of Science.

Braced Oil Dampers for Buildings

Development and Analytical Modeling

Osamu Takahashi and Atsuki Yokoyama

CRC Press
Taylor & Francis Group
Boca Raton London New York

CRC Press is an imprint of the
Taylor & Francis Group, an **informa** business

First edition published 2024
by CRC Press
6000 Broken Sound Parkway NW, Suite 300, Boca Raton, FL 33487-2742

and by CRC Press
4 Park Square, Milton Park, Abingdon, Oxon, OX14 4RN

CRC Press is an imprint of Taylor & Francis Group, LLC

Library of Congress Cataloging-in-Publication Data

Names: Takahashi, Osamu (Researcher in earthquake engineering), author. | Yokoyama, Atsuki, author.
Title: Braced oil dampers for buildings : development and analytical modeling / Osamu Takahashi and Atsuki Yokoyama.
Description: First edition. | Boca Raton, FL : CRC Press, 2024. | Includes bibliographical references and index.
Identifiers: LCCN 2023010169 | ISBN 9781032268637 (hbk) | ISBN 9781032268644 (pbk) | ISBN 9781003290261 (ebk)
Subjects: LCSH: Buildings--Earthquake effects. | Earthquake resistant design. | Damping (Mechanics) | Seismic waves--Damping. | Shock absorbers.
Classification: LCC TH1095 .T348 2024 | DDC 693.8/52--dc23/eng/20230406
LC record available at https://lccn.loc.gov/2023010169

ISBN: 978-1-032-26863-7 (hbk)
ISBN: 978-1-032-26864-4 (pbk)
ISBN: 978-1-003-29026-1 (ebk)

DOI: 10.1201/9781003290261

Typeset in Sabon
by Deanta Global Publishing Services, Chennai, India

Access the Support Material: www.routledge.com/9781032268637

Contents

6 Analytical model and verification of building oil damper under small amplitude   103

7 Conclusion to Part 1   125

Part 2
Development of the oil damper stiffness for architectural vibration control and experimental research on structural characterization

Part I

Development and analytical modeling of braced oil dampers for buildings

Chapter 1

Introduction

1.1 BACKGROUND OF THE RESEARCH

Japan is one of the most earthquake-prone countries in the world and has been frequently damaged by large earthquakes since ancient times. According to the White Paper on Disaster Management,[1] as shown in Figure 1.1, the number of earthquakes in Japan with a magnitude of 6.0 or above was 95 in the 5 years from 1994 to 1998, which shows that about 20% of the total number of earthquakes worldwide occurred in Japan. In addition, human damage has also occurred due to the earthquake; the 1995 Hyogo-ken Nanbu Earthquake claimed more than 6,000 lives and the gross national product decreased by approximately 2.0–3.0%, as shown in Figure 1.2. Earthquakes are a familiar natural disaster for people living in Japan, who have suffered from the damage caused by large earthquakes since ancient times, and it has long been an important concern for them to ensure the safety of the buildings they live in.

Most Japanese buildings were wooden buildings that utilized abundant forest resources. Although Horyuji Temple, the oldest wooden building in the world, still exists, most of the buildings before the Edo period were wooden and they were supposed to be rebuilt every time they were damaged, except for buildings such as storehouses.

Table 1.1 shows the timeline about seismic structure in Japan. Since the Meiji Restoration, brick and masonry building techniques from Western Europe with excellent fire resistance have been introduced. However, the Nobi Earthquake of 1891 caused great damage not only to wooden structures but also to brick and masonry buildings, which raised questions over the reliability of those structures. In the following year, 1892, the Earthquake Investigation Committee was established, and it can be said that research and design of seismic structures began in earnest.

The Great Kanto Earthquake of 1923 caused great damage, mainly in Tokyo, but there was little damage to the former Nihon Kogyo Building, which has a rigid seismic structure with seismic walls, and other buildings that have undergone seismic calculations. They demonstrated the effectiveness of the rigid seismic structure. In 1924, the Urban Building Law

DOI: 10.1201/9781003290261-2

Earthquakes of M6.0 or greater

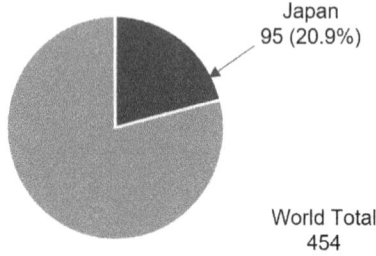

Japan
95 (20.9%)

World Total
454

※ Sum of 1994 - 1998

Figure 1.1 Earthquake frequency statistics.

(JP¥ 1 trillion)

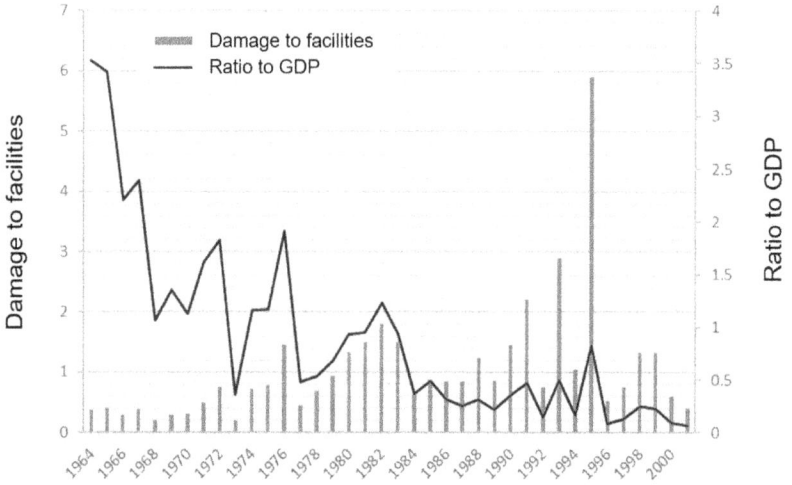

* Prepared by the Cabinet Office based on data from various ministries and agencies.

Figure 1.2 Earthquake damage statistics.

Enforcement Regulations were significantly revised; these were the first seismic design regulations introduced to add 10% of the total weight of a building (design seismic intensity 0.1) to the building as the seismic force.

After that, based on the idea of the dynamic behavior of the building at the time of an earthquake, the idea of a flexible structure was proposed, in which it is better to lengthen the natural period in a building with low rigidity (flexible structure) to reduce the influence of input seismic motion. Due to the difference from the idea of a rigid structure that it is better to increase the yield strength, a controversy over this topic was raised, making research and the designing of trials active in a good sense.

Table 1.1 History of seismic structures[2]

Year	Affair
1891	• Nobi Earthquake
1892	• Earthquake Prevention Research Committee
1916	• Seismic structure theory (SANO Toshikata)
	(Seismic intensity method)
1922	• Theory of earthquake-resistant structure of construction (NAITO Tachu)
	(Effects of earthquake-resistant walls, seismic calculation method - Former Kogyo Bank of Japan)
1923	• Great Kanto Earthquake
	• Damage to U.S. and European direct import buildings
1924	• Designed seismic intensity 0.1 for urban buildings
	• Height 100 shaku (31m) limit
After	• Popularization of RC (Reinforced Concrete) and SRC (Steel Reinforced Concrete)
	• Stress calculation method of building construction
	• Vibrational research
	[Soft Rigid Controversy]
	• Development of strong motion meter → Analysis of actual earthquakes
	• Computer development → motion response of structures
	• Consideration for high-rise Tokyo station
	• Analysis method (mainly nonlinear)
	• Rigidity distribution
1963	• Elimination of height restrictions
	• Revision of the Building Standards Act (Article 39 of Minister Special Recognition)
1968	• Kasumigaseki Building
1981	• New seismic design method

After that, large earthquakes such as the Tonankai Earthquake (1944), Mikawa Earthquake (1945), Nankai Earthquake (1946), and Fukui Earthquake (1948) occurred frequently and caused great damage. Thereby, in 1950, the Building Standards Law was enacted, and it was legally decided to adopt a value of 20% of the total weight of the building (design seismic intensity 0.2) as the seismic force in the structural regulations of the law enforcement order.

From the latter half of the 1950s, the reconstruction of the Japanese economy began in earnest, and the demand for advanced utilization of urban space increased. The Specific Block System and limiting the floor-area ratio were introduced, respectively, in 1961 and 1963. The building height limit was lifted with the revision of the Building Standards Law (1965), and the environment that enables the construction of high-rise buildings was prepared. Around this time, the development of computers, seismic response analysis technology, and programs for structures based on earthquake records progressed, and in 1968, Japan's first skyscraper, the "Mitsui Kasumigaseki Building," was constructed; skyscrapers in earthquake-prone countries became feasible.

The Niigata Earthquake (1964) and the Tokachi-oki Earthquake (1968) caused great damage due to liquefaction of sandy ground, shear damage on reinforced concrete short columns, and collapse of buildings with remarkable eccentricity. Due to the revision of the Building Standard Act Enforcement Order in 1971, structural regulations such as the spacing between strips have been strengthened. In addition, the 1978 Miyagi-ken-oki Earthquake triggered a major revision of the Building Standards Law in 1981, and the New Seismic Design Law was adopted as a new regulation. The main revisions of the new seismic design method are (1) dynamic consideration for seismic force regulations, (2) restriction of inter-story deformation angle, and (3) confirmation of possessed horizontal strength, which was a conventional design method based on constant seismic intensity. However, in the New Seismic Design Method, the seismic force is determined by the ground type and the natural period, and in addition to the idea based on the dynamic behavior of the building, which has a larger acting force on the upper floors, it targets large earthquakes. The design seismic intensity was set to 1.0, and it was required to prevent the collapse of buildings and ensure the safety of human life, depending on the energy absorption capacity.

The validity of this new seismic design method was proved necessary by the extensive damage to buildings in the Hyogo-ken Nanbu Earthquake (1995), but on the other hand, the following questions were also raised.

(1) An essential question about the idea that human life is protected but the function and property of the building do not have to be guaranteed.
(2) Higher demand for earthquake safety for buildings that are important for disaster prevention.
(3) Request for the asset value of the building and the retention of the owned property.

Technological research and development are continuing to meet the social needs that the current seismic structures may have limitations in responding to these problems. With the social demand for safer and more comfortable buildings, the researchers' target has shifted from seismic resistant structure, which simply resists shaking, to seismic isolation structure, which separates the ground from buildings, and vibration control, which absorbs and controls shaking.

Table 1.2 shows the classification of seismic resistance, seismic isolation, and vibration control structure in the anti-seismic structure organized in this paper.

Table 1.2 Classification of seismic resistance, seismic isolation, and vibration control structure in anti-seismic structure[3]

Structure	Structural format/ type	Passive, active	Components and systems
Earthquake-resistant structure	Rigid structure/ Strength type	Passive	• RC earthquake-resistant wall • Steel brace
	Soft structure/ Toughness type	Passive	• Ramen structure
Seismic isolation structure	Non-resonant structure	Passive	• Natural rubber-based laminated rubber • High damping rubber-based laminated rubber • Laminated rubber with lead plug • Sliding support • Lead damper • Steel bar damper • Oil damper for seismic isolation
		Semi-active type	Semi-active oil damper
Damping structure	Mass addition type	Passive Hybrid	• TMD[*1] • TLD[*2] • HMD[*3] • APTMD[*4]
		Active type	• AMD[*5]
	Energy absorption mechanism type	Passive	• Steel dampers • Viscous damper • Viscoelastic damper • Oil damper
		Semi-active type	• Semi-active oil damper • AVS[*6] • AVD[*7]

[*1] Tuned Mass Damper
[*2] Tuned Liquid Damper
[*3] Hybrid Mass Damper
[*4] Active-Passive Tuned Mass Damper
[*5] Active Mass Damper
[*6] Active Variable Stiffness
[*7] Active Variable Damping

The concept of seismic structures is the structure of a building (columns, beams, walls, braces), which converts external force caused by seismic motion or the like into strain energy and absorbs it. Therefore, the structural form can be divided into two types. One is a type in which the structure is deformed, and yield hinges are formed in the columns and beams to absorb energy while increasing the horizontal displacement (flexible structure, toughness type), and the other is the entire structure. The columns/beams or seismic walls are of the type (rigid structure, strength type) that absorbs large energy with small deformation by increasing rigidity and strength. All seismic structures aim to prevent damage to the structure in medium earthquakes but prioritize the protection of human life in the event of a large earthquake, and it is unavoidable that the building will be damaged. Careful consideration and expense may be required when the damaged buildings are going to be used after the disaster.

The concept of seismic isolation structure is a non-resonant structure by vibration insulation, and various attempts to insulate buildings from seismic motion have been proposed since around 1890 in Japan. The seismic isolation structure was put into practical use in Japan in the 1980s, and the construction of the Yachiyodai residence in Chiba prefecture (1983, Hideyuki Tada, Shoichi Yamaguchi) was the first. After the Hyogoken-Nanbu Earthquake in 1995, the structure's superiority was demonstrated, and the number of application cases increased sharply. In a seismic isolation structure, a seismic isolation layer composed of seismic isolation members is generally provided between the ground and the building, and the natural period of the structure system is set to a long period of about 2 to 3 sec. due to the soft spring characteristics of the seismic isolation materials. It is a structure that reduces the transmission of energy due to seismic isolation to the superstructure by adding a large damping force by setting it to and installing a damper. This makes it possible to reduce the response acceleration of the building to about 1/3 to 1/4 of the input acceleration and prevent not only damage to the building but also secondary damage due to movement or the falling of equipment and stored items.

The concept of a vibration-damping structure is the idea of using various dampers to suppress shaking due to disturbance energy such as seismic motion, and the type that adds mass and absorbs energy (mass addition type) and energy in the building frame. There is a type that incorporates an absorption mechanism (energy absorption mechanism type). The mass-added type is expected to have a vibration-damping effect on the primary natural period of the building and is often used as a countermeasure against the wind rather than earthquakes. In this paper, we will focus on the energy absorption mechanism type as a vibration-damping structure that is effective against earthquakes. Further, regarding the vibration-damping member used in the energy absorption mechanism type, the idea has been used for a long time in the field of mechanical engineering, for example, railroad vehicles and automobiles. In railway vehicles, leaf springs have been used in the past, and oil dampers have recently been used to suppress vehicle vibration. The latest Shinkansen 500 series and 700 series vehicles detect the movement of the vehicle and use a computer-controlled oil damper, that is, a semi-active oil damper, to control vibration during the high-speed operation of the Shinkansen. Even in automobiles, oil dampers are used in parts such as the vehicle chassis to achieve a comfortable driving performance.

After the Hyogoken-Nanbu Earthquake in 1995, the safety myth of conventional seismic structures collapsed, and the development of seismic isolation structures and vibration control structures as one effective idea of new anti-earthquake structures began to attract social attention. It is thought that the vibration control structure is expected to be used in a wide range of buildings because it is not easily affected by the restrictions of the ground and site like the seismic isolation structure and can be adopted even in buildings that are vulnerable to falls.

Figure 1.3 shows structure of the building and the human body. If the concept of seismic structure and vibration control structure is compared to the human body structure, the seismic structure can protect human life while destroying the building. It can be said that it is a structure that tries to protect important objects such as internal organs by absorbing disturbance energy while destroying the bones of the human body. On the other hand, in terms of absorbing energy with a damper added by the target of the vibration-damping structure and protecting the building without damage, an energy absorption mechanism such as muscle is incorporated in the bone to leave the bone intact and disturbing energy. It can be said that it is a structure that absorbs energy with muscles.

With this background, the author began researching damping structures around 1996, believing that the need for damping structures would continue to grow.

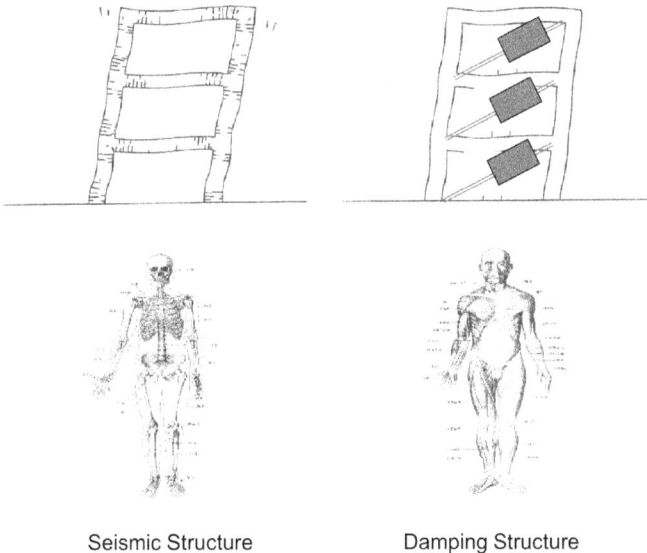

Seismic Structure   Damping Structure

Figure 1.3 Concepts of seismic and damping structures.

1.2 ABOUT THE VIBRATION CONTROL SYSTEM

At 2006 (Wrote the Japanese version of this paper), a vibration control system incorporating various vibration control members has been adopted as a building with a vibration control structure. The uses, structural types, and scales of vibration-damping buildings are wide ranging and are being applied in most cases in steel-framed high-rise buildings.

Before the 1995 Hyogoken-Nanbu Earthquake, many of the control objectives were aimed at reducing the wind sway response, but in recent years, many have also been aimed at reducing the seismic response. Steel structures are often found by structural type, but the scope has been expanded to include reinforced concrete structures and wooden detached houses. In addition, the number of cases where vibration control structures are being adopted not only for new buildings but also for seismic retrofitting of existing buildings is increasing.

1.2.1 Damping components

Table 1.3 shows the types of damping members that are generally used. The types of vibration-damping members are broadly classified into four types: oil dampers, viscous dampers, viscoelastic dampers, and steel dampers.

> F: Damping resistance
> $K(\omega)$: Rigidity
> $f(D)$: Displacement-dependent function
> V: Velocity
> $C(\omega)$: Attenuation coefficient
> ω: Circular frequency
> D: Displacement

Table 1.4 shows a list of the basic characteristics of each damping member.

While steel dampers are displacement-dependent history damping types, oil dampers, viscous dampers, and viscoelastic dampers are speed-dependent viscous damping types. In addition, viscous dampers, viscoelastic dampers, and steel dampers have material-dependent damping resistance, while oil dampers are machining damping mechanisms that utilize the turbulent flow resistance of oil as damping resistance is different.

The damping effect also differs depending on such differences in basic characteristics.

Since the steel damper is a displacement-dependent type, it cannot be used because there is a concern about fatigue due to repetition at the wind habitability level that frequently occurs. Care must be taken when using it even at small and medium-sized earthquake levels with a seismic intensity of about 3. Even with the same speed-dependent vibration-damping

Table 1.3 Types of dampers[4]

	Oil Damper	Viscous Damper	Viscoelastic Damper	Steel Damper
F :	C * V	C * V$^\alpha$	K(ω) * D + C(ω) * V	K * f(D)
F-D Curve :	Ellipse	Ellipse + Rectangle	Tilted Ellipse	Bi-Linear

member, the viscous damper uses a highly viscous fluid as the material, so the viscous fluid is fixed at a minute level below the wind habitability level and does not generate viscous resistance. Therefore, it is necessary to be careful when using it.

In addition, since the materials of the viscous damper and the viscoelastic damper are temperature-dependent, the vibration-damping effect is also affected by the temperature fluctuation of the vibration-damping member; therefore it is necessary to fully consider the fluctuation. This effect also adversely affects performance reproducibility.

Regarding the restrictions on the plan, there are cases where the viscous damper cannot be placed on the plan because it is a vibration-damping member with a large wall shape in the elevation. It should be noted that the viscous damper may cause the high-viscosity viscous fluid inside to flow out if it is tilted or tilted sideways in the construction plan.

Table 1.4 lists the advantages and disadvantages of each damping member.

According to this list, as a damping member, the oil damper can be expected to have a stable damping effect from small amplitude to large amplitude compared to other damping members and is few affected by temperature dependence and performance deterioration due to fatigue.

Considering these characteristics and advantages/disadvantages, the oil damper is superior to other vibration-damping members, and the author started research on the vibration-damping structure using the oil damper from around 1996.[5]

1.2.2 Structural form of damping components

As shown in Table 1.5, the structural types of vibration-damping members in practical use (2006) can be classified into three types: direct bonding type, indirect bonding type, and other types.

Table 1.4 Advantages and disadvantages of dampers

		Oil damper	Viscous Damper	Viscoelastic Damper	Steel damper
Basic characteristics	Damping type	Viscous damping type	Viscous damping type	Viscous damping type	Historical damping type
	Damping characteristics	Speed-dependent type	Speed-dependent type	Speed-dependent type (+ elastic type)	Displacement-dependent type
	Damping mechanism	Machining	Material dependent	Material dependent	Material dependent
	Damping resistance	Turbulent resistance	Viscous resistance	Shear resistance	Plasticity history
	Material	Oil	Polymeric compounds	Acrylic compounds Diene compounds	Steel
	Shape	Cylindrical	Cylindrical Wall type	Cylindrical Face type	Square Panel type
Damping effect	Wind residence level	○	△	○	×
	Small and medium earthquake level	◎	◎	◎	△
	Rare earthquake level	◎	◎	◎	○
	Extremely rare earthquake level	◎	◎	◎	◎
Cost	Equipment cost	○	○	△	◎
	Reduced frame cost	○	○	○	○
Constraint	Floor plan	△	×	△	△
	On construction plan	○	△	○	○
Achievement	Building field	○	○	△	○
	Machinery field	◎	△	△	×

(Continued)

Table 1.4 Continued Advantages and disadvantages of dampers

		Oil damper	Viscous Damper	Viscoelastic Damper	Steel damper
Independence	Manufacturer-dependent	○	○	△	○
	Limited to manufacturers	○	△	×	○
Other (various characteristics)	Temperature dependence	○	×	×	○
	Fatigue characteristics	○	○	△	×
	Repeatability	○	○	△	×
	Performance reproducibility	○	×	×	○
	Maintenance-free	○	○	○	○

The direct joining type is a type in which the vibration-damping member is directly connected to the main frame between the layers. Specifically, there are brace types, wall types, and shear-link types. The indirect joint type transmits the interlayer deformation through bending deformation of a beam, a bundle, or the like. Specifically, there are stud types, square cane types, and joint types. Other than that, it shall have a mechanism that utilizes the total deformation of the main frame and a mechanism that amplifies the inter-story deformation. Specifically, there are pillar types, outrigger types, amplification mechanism types, and the like.

Of these, the structural type that uses an oil damper as a vibration-damping member in the direct joint type brace is the result of the structural type that the author studied, focusing on the oil damper from around 1996 and the research and development process is described in Section 1.4.

Table 1.5. Structural type classification of dampers

Table 1.5(a) Combination of four damping types and structural types of dampers (1) (4)

	Brace Type	Wall Type	Shear Link Type	Stud Type
Oil Damper				
Viscous Damper				
Viscoelastic Damper				
Steel Damper				

Table 1.5(b) Combination of four damping types and structural types of dampers (2) (4)

	Bracket Type	Connector Type	Column Type	Others
Oil Damper				oil / outrigger
Viscous Damper				oil / amplifier
Viscoelastic Damper				oil + history / shear link
Steel Damper		friction		stud / friction / etc.

1.3 PREVIOUS STUDIES

The vibration-damping structure is a structure that effectively absorbs input energy by installing a member with high energy absorption performance in the main structural part of the building and makes it possible to reduce damage to the main structural part. It is a structure that not only ensures the safety of the building at the time but also reduces the shaking caused by small and medium-sized earthquakes and wind.

The idea of using a damping member to control the response of a building's disturbance energy has long been considered: "Proposal to use wire by pretension" (1954, Takuji Kobori) and "Vibration experiment by friction damper" (1957, Yasutoshi Sonobe). Viscoelastic dampers were used at the World Trade Center in New York at the end of the 1960s for use as a damping structure for buildings. In Japan, Kiyoo Matsushita and Masatetsu Izumi have adopted a double-column structure using a reinforcing bar damper in the No. 1 building of Tokyo Science University to realize a system that is a pioneer in the vibration control structure that ensures safety in the event of a large earthquake.

The following is an outline of past research examples of each damping member. The affiliation is the affiliation at the time of research presentation.

As an example of research on oil dampers, such research was carried out mainly by Kashima Construction/Kobori Laboratory, and in 1994, Niwa (Kashima Construction/Kobori Laboratory) et al.[6] had an oil reservoir above the oil damper. We have reported an oil reservoir external type oil damper (referred to as a conventional oil damper) that can realize a large capacity and a stable high damping coefficient regardless of temperature by arranging the oil reservoir. In 1996, using the conventional oil damper developed in this report, Kurino (Kajima Construction/Kobori Laboratory) et al.[7, 8, 9] applied it to a 26-story steel-framed skyscraper. An example and analytical modeling are reported. At this time, the conventional oil damper proposes modeling with the Maxwell model, but the damping coefficient C is treated as the equivalent linear Maxwell model without considering the nonlinearity. After that, in 1998, the author (Structural Planning Institute) built an oil reservoir inside the piston rod so that it could be installed in a building in a brace shape. After publishing the report, the solution method using the nonlinear Maxwell model, and the application example in a 9-story steel-framed building as the first practical application in the world, the authors developed it as a vibration control system for architecture. It can be said that the number of cases where oil dampers are used has increased rapidly and has reached the present. Examples include Hanamura (Obayashi) et al.[10, 11] reporting its application to a 43-story skyscraper in 2001, and Asai (Konoike Construction) et al. [12, 13, 14] in 2002 reporting the application of a 35-story CFT (Concrete Filled Steel Tube) structure to a skyscraper.

As an example of research on viscous dampers, we report an example in which research on viscous vibration-damping walls using viscous fluid has been carried out mainly by Miyazaki (Sumitomo Mitsui) et al.[15] since 1992. In 1998, Sera (Sumitomo Construction) et al.[16, 17] reported an application example of this viscous damping wall in a 24-story skyscraper, and in 1999, Nakano (Postal Minister's Office, Facility Department) et al.[18, 19, 20, 21, 22] reported application examples in a 28-story steel-framed government building, performance confirmation tests during construction, and results of human-powered vibration tests. Moreover, there are several reports regarding viscous dampers by overseas makers as well.[23, 24]

As examples of research on viscoelastic dampers, there are many cases in which research results are reported in the form of joint research with university institutions, centered on researchers from manufacturers of viscoelastic bodies. In 1996, Soda (Waseda University) et al. and Ishikawa (Showa Densen Denki) et al.[25] reported the basic mechanical properties of three types of viscoelastic bodies: polynorbornene-based, butyl rubber-based, and silicon rubber-based. Regarding these analytical models, Takahashi (Waseda University) et al.[29, 30] reported the analysis study by calling the model in which the Maxwell model is arranged in parallel with three elements the M3 model. In 2001, Kasai (Tokyo Institute of Technology) et al. and Okuma (Sumitomo Styrene) et al.[31, 32] reported an analysis study of an acrylic viscoelastic body using a Kelvin model, and in 2002, Sone (Yokohama Rubber) et al. and Nakamura (Shimizu Construction) et al.[33, 34, 35] called an analysis model for viscoelastic body using a styrene-based elastomer a four-element model with a spring, Maxwell model, and dashpot in parallel and reported accuracy of the model. They also reported how to set up a four-element model for time history response analysis using a building model. In 1997, Mori (Konoike Construction) et al.[26, 27, 28] reported an application example in a 40-story CFT-built skyscraper. The viscoelastic body at this time was made by Showa Densen Denki, and the analysis model at the time of building design was reported using the Voigt model.

As examples of research on steel dampers,[36–44] many studies have been reported by steel makers, general contractors, and researchers at universities. Many steel dampers have been developed with ultra-low yield point steel that can absorb energy from a lower stress level than commonly used construction steel and have extended deformation performance and are damping dampers. It is used as a brace type, wall type, shear-link type, and stud type.

1.4 PURPOSE OF THIS BOOK

Since around 1996, the author has realized the social necessity of vibration control structures instead of seismic structures, and after comparing the characteristics of various vibration control members, he focused on the superiority of oil dampers and used oil dampers for vibration control.

In the conventional oil dampers in the previous studies, the oil reservoir was placed above the cylinder of the oil damper. Therefore, there are restrictions on the installation: the oil damper can be installed only horizontally and the installation method is restricted to the use of the shear-link type. The following three points can be said to be problems in the shear-link type installation.

(1) Peripheral members need to be reinforced because the damping force adversely affects the structural frame.
(2) It is not easy to remove during construction.
(3) A guide is required at the link part to prevent deformation in the out-of-plane direction.

To solve these problems, it is effective to install the oil damper part in series in a brace shape and install it as a brace-type oil damper instead of mounting it with a shear-link type as in the past. The author thought about it and started research from around 1997 with the main purpose of developing a brace-type oil damper considering damping characteristics and various dependencies and researching an analysis model.

The following also shows the purposes considered while the main purpose was to develop a brace-type oil damper.

(1) The purpose is to develop a building oil damper with a structure that can be installed freely up and down, horizontally, and diagonally.
(2) Since it cannot be said that the performance confirmation regarding various dependencies (speed dependence, temperature dependence, temperature rise, durability) is sufficient for the conventional oil damper, these are systematically confirmed and organized in the construction oil damper.
(3) Since the conventional oil damper handles the equivalent linear Maxwell model as an analysis model, the accuracy of the analysis value is a problem, and the purpose is to treat it as a nonlinear Maxwell model and propose a solution.
(4) Install brace-type oil dampers in the design and construction of actual buildings.
(5) Check the damping effect of the brace-type oil damper in the actual building.
(6) The purpose is to confirm the damping effect and build an analysis model even in the micro-vibration region where there are few examples in the mechanical field.
(7) The purpose of the end is to develop a method that can be joined on-site and can be directly attached to the building structure in consideration of workability at the site.

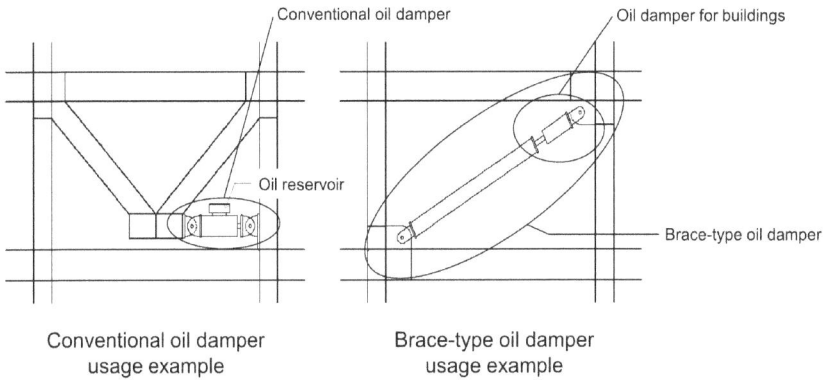

Figure 1.4 Usage examples of oil dampers.

Figure 1.4 shows an example of using a conventional oil damper and an example of using a brace-type oil damper when using a building oil damper targeted in this study. The brace-type oil damper is basically a structure in which an oil damper portion and a steel pipe braces portion are connected in series.

1.5 STRUCTURE OF THIS BOOK

This book consists of seven chapters. The outline of each chapter is shown below.

In Chapter 1, "Introduction," the necessity of the damping structure is described as the background of the research, and the damping members are organized to focus on the superiority of oil dampers. The findings and problems obtained from previous studies on damping structures are organized, and the significance and positioning of this paper are clarified.

In Chapter 2, "Development of building oil dampers considering damping characteristics and various dependencies," the development of new building oil dampers that can be attached to the brace shape in consideration of damping characteristics and various dependencies is discussed. Here, the structural outline and hydraulic circuit diagram of the building oil damper are shown, and the points developed by focusing on the new building oil damper are clarified. The unit experiments of building oil dampers are systematically organized in consideration of damping characteristics and various dependent characteristics. As experiments for confirming various dependent characteristics, temperature dependence confirmation experiments, temperature rise confirmation experiments, durability confirmation experiments, and random wave response confirmation experiments were conducted, and the dependent characteristics were determined according to the temperature, speed, and frequency. This is systematically organized and shown.

In Chapter 3, "Analysis model of a single building oil damper," a nonlinear Maxwell model is presented as an analysis model of a newly developed building oil damper that can be attached to a brace shape, and a numerical calculation algorithm is proposed. Quantitative consistency is examined by comparison with various experiments conducted in Chapter 2, and problems with analysis parameters are clarified.

In Chapter 4, "Damping characteristics and analysis model of brace-type oil damper," a brace-type oil damper is manufactured with the actual length assuming it will be installed in a real building, and the damping characteristics and analysis model are reviewed. A full-scale dynamic vibration experiment or a full-scale frame dynamic vibration experiment is carried out using an oil damper, and the consistency is examined and quantitatively evaluated by comparing the experimental results with the proposed analysis model. It shows that it can be done. In addition, safety and workability are confirmed at the actual scale level, and the possibility of practical application is verified.

In Chapter 5, "Vibration control performance evaluation and confirmation experiment in an actual building using a brace-type oil damper," a structural example of a building using a brace-type oil damper is shown, and this actual building is shown. The confirmation of the damping effect by the vibration-damping experiment, the human-powered vibration experiment, and the long-term observation conducted in the shaker experiment, in particular, are compared both with and without the brace-type oil damper in the experiment. It is clearly shown that the brace-type oil damper exerts a vibration-damping effect by absorbing energy from a minute area in small and medium-sized earthquakes.

Chapter 6, "Analysis model and verification of building oil damper under minute amplitude," shows that it is necessary to propose and examine a more detailed analysis model for the analysis model of building oil dampers under minute amplitude such as environmental vibration. The analysis model is proposed, and the quantitative consistency is clearly shown by comparison with the actual measurement results as an application example of the minute amplitude model. When proposing a more detailed analysis model, we focus on the sliding resistance inside the building oil damper and the mixing rate of air bubbles inside the oil.

Chapter 7, "Conclusions," summarizes this study and presents the future potential of brace-type oil dampers. As a future possibility for the brace-type oil damper, we are proposing an improved oil damper that has a restoring force characteristic by improving the internal mechanism.

As an appendix, examples of a skyscraper and a reinforced building are shown as practical examples in actual buildings using a brace-type oil damper. Regarding administrative handling in the structural design of actual buildings using the brace-type oil damper, it is necessary to examine the structures by the time history route. When reducing the seismic force for design in a new building from the value specified by the new seismic design law, a ministerial approval evaluation is required, and even if the seismic force for design is not reduced, it is voluntary by a third party in principle. Structural evaluation is required. In addition, there are many cases where a third-party organization evaluates a reinforced building.

REFERENCES

1. Cabinet Office: White paper on disaster prevention, Printing Bureau, Ministry of Finance, published on July 23, 2001.
2. Takuji Kobori: *Seismic Control Structure Theory and Practice*, Kajima Institute Publishing, 1993.9.
3. Japan Structural Consultants Association: *Response Control Structure Design Method*, Shokokusha, 2000.11.
4. Japan Seismic Isolation Structure Association: *Passive Vibration Control Structure Design/Construction Manual*, Meibunsha, 2004.10.
5. Postal Services Agency Facility Information Department: *Seismic Isolation and Vibration Control Building of Postal Services*, Facility/Construction Information Center, 2002.3.

[REFERENCES ON OIL DAMPERS] FOR EXAMPLE

6. N. Niwa etd.: "Passive seismic response control system with high performance oil damper," *10th European Conference on Earthquake Engineering*, Vienna, 1994.
7. Kazuo Ishihara, Takuji Kobori, Yoshiaki Suzuki, Kazufumi Horiuchi, Koichi Harashima: "Damping of high-rise buildings with high-damping oil dampers (1) outline of applicable buildings," *Architectural Institute of Japan Academic Lecture Summary*, B-2 Volume, pp.851, 1996.
8. Atsushi Tanaka, Koji Ishii, Katsuya Igarashi, Motoichi Takahashi, Naomi Niwa: "Damping of high-rise buildings with high damping oil dampers (Part 2) vibration experiment with pendulum," *Architectural Institute of Japan Academic Lecture Summary*, B-2 Volume, pp.853, 1996.
9. Haruhiko Kurino, Genichi Takahashi, Naomi Niwa: "Damping of high-rise buildings with high-damping oil dampers (3) examination by simulation analysis," *Architectural Institute of Japan Academic Lecture Summary*, B-2 Volume, pp.855, 1996.
10. Hirotsugu Hanamura, Yasuhiro Kanzaka, Hiroyuki Ota, Masayuki Yamanaka, Takeshi Sano: "(Tentative) Dentsu new building construction project- (10) oil damper performance test results part 1," *Abstracts of Academic Lectures at the Japan Building Convention*, B-2 Separate volume, pp.387, 2001.

11. Yasuhiro Kanzaka, Hirotsugu Hanamura, Hiroyuki Ota, Masayuki Yamanaka, Takeshi Sano: "(Tentative) Dentsu new building construction project- (part 11) oil damper performance test results part 2," *Abstracts of Academic Lectures at the Japan Building Convention*, B-2 Separate volume, pp.389, 2001.

12. Jun Asai, Kazuo Kondo, Kenichi Kashihara: "CFT skyscraper using oil damper part 1. Building overview and oil damper," *Abstracts of Academic Lectures of the Japan Building Convention*, B-2 Volume, pp.703, 2002.

13. Naoto Tani, Yasuo Kuroki, Nozomi Ikawa: "CFT skyscraper using oil damper part 2. Seismic response analysis," *Abstracts of Academic Lectures at the Japan Building Conference*, B-2 Volume, pp.705, 2002.

14. Shinji Yoshikawa, Hiroshi Ota, Shinji Ito: "CFT skyscraper using oil damper part 3. Examination of wind load," *Abstracts of Academic Lectures of the Japan Building Convention*, B-2 Volume, pp.707, 2002.

[REFERENCES ON VISCOUS DAMPERS] FOR EXAMPLE

15. Mitsuo Miyazaki, Yuji Mitsusaka, Kokukuni Kato, Toyokatsu Takahashi, Kazunori Mizuto, Toshio Sakakibara: "High-rise building using viscous damping wall-structural design of TV Shizuoka media city building," *Building Letter*, pp.1, 1992.2.

16. Shinji Sera, Noboru Hayakawa, Mitsuo Miyazaki, Yukihiro Nishimura: "Design of high-rise buildings using viscous damping seismic walls (Part 1: Design of seismic control structures)," *Architectural Institute of Japan Academic Lecture Abstracts*, B -2 Separate volumes, pp.879, 1998.

17. Shinji Sera, Noboru Hayakawa, Hiroshi Ogura: "Design of high-rise buildings using viscous damping wall (Part 2: Performance test of viscous damping wall)", *Architectural Institute of Japan Academic Lecture Summary*, B-2 Separate volume, pp.989, 1999.

18. Tsuyoshi Nakano, Taichi Yoshida, Shiro Miuchi, Hisaya Tanaka: "Seismic control structure using viscous damping wall (1) performance confirmation test of viscous damping wall-," *Architectural Institute of Japan Academic Lecture Summary*, C-1 Separate volume, pp.869, 1999

19. Koumori Ryokawa, Hisaya Tanaka, Osamu Takahashi: "Damping structure using viscous damping wall (Part 2) consideration on seismic analysis model in micro-amplitude region," *Architectural Institute of Japan Academic Lecture Summary*, C-1 Separate volume, pp.871, 1999.

20. Tsuyoshi Nakano, Ryoichi Yajima: "Damping structure using viscous damping wall (Part 3) vibration characteristics test and observation plan," *Architectural Institute of Japan Academic Lecture Summary*, B-2 Volume, pp.1057, 2000.

21. Masayoshi Kuno: "Damping structure using viscous damping wall (Part 4) constant microtremor measurement and human power vibration," *Architectural Institute of Japan Academic Lecture Abstracts*, B-2 Volume, pp.1059, 2000.

22. Yoko Hirakawa, Toshifumi Okuzono, Osamu Takahashi: "Damping structure using viscous damping wall (Part 5) verification of analytical model and experimental results," *Architectural Institute of Japan Academic Lecture Summary*, B-2 Volume, pp.1061, 2000.

23. Michael C Constantinou, MD Symans, Panos Tsopeals, DP Taylor: "Fluid viscous dampers in application for seismic energy dissipation and seismic isolation," Proc. ATC-17-1 Seminar on Seismic Isolation, Passive Energy Dissipation, and Active Control, San Francisco, March 2, pp.581–592, 1993.
24. Michael C Constantinou, MD Symans: "Experimental and analytical investigation of seismic response of structure with supplemental fluid viscous dampers," Report # NCEER-92-0032, National Center for Earthquake Engineering Research, SUNY/Buffalo, Buffalo, New York, 1992.

[REFERENCES ON VISCOELASTIC DAMPERS]
FOR EXAMPLE

25. Kazuhisa Ishikawa, Katsuhiro Teshiroki, Mayuya Soda, Masaru Oishi: "Study on viscoelastic damper part 1 basic mechanical properties of viscoelastic body materials," *Abstracts of Academic Lectures of the Japan Building Conference*, B-2 Volume, pp.873, 1996.
26. Hiroshige Mori, Kenichi Kashihara, Junichiro Ishida, Yasuo Kuroki: "CFT super high rise housing using viscoelastic dampers (1) outline of architecture and viscoelastic dampers," *Abstracts of Academic Lectures at the Japan Building Convention*, B-2 Volume, pp.907, 1997.
27. Jun Asai, Kenichi Kashihara, Yasuo Kuroki, Hiroshige Mori: "CFT-built skyscraper using viscoelastic damper (2) wind response analysis (examination of habitability)," *Architectural Institute of Japan Academic Lecture Summary*, B-2 Separate volumes, pp.909, 1997.
28. Shigeru Fujiuchi, Kenichi Kashihara, Yasuo Kuroki, Hiroshige Mori: "CFT-built super high-rise housing using viscoelastic dampers (3) seismic response analysis," *Abstracts of Academic Lectures at the Japan Building Conference*, B-2 Volume, pp.911, 1997.
29. Yuji Takahashi, Mayuya Soda: "How to determine the installation capacity of a viscoelastic damper in a building structure (Part 1. Formulation using a mass elastic model)," *Abstracts of Academic Lectures at the Japan Building Conference*, B-2 Volume, pp.867, 1998.
30. Mayuya Soda, Yuji Takahashi: "How to determine the installed capacity of a viscoelastic damper in a building structure (Part 2. Consideration of the seismic response spectrum of the model)," *Abstracts of Academic Lectures of the Japan Building Conference*, B-2 Volume, pp.869, 1998.
31. Kazuhiko Kasai, Kiyoshi Okuma: "Proposal of simple modeling of linear viscoelastic damper by Kelvin body-part 1: Theory and accuracy verification using constant vibration," *Abstracts of Academic Lectures of the Japan Architecture Conference*, B-2 Volume, pp.393, 2001.
32. Kiyoshi Okuma, Kazuhiko Kasai: "Proposal of simple modeling of linear viscoelastic damper by Kelvin body-part 2: Time history analysis and accuracy verification using random wave input," *Abstracts of Academic Lectures of the Japan Architecture Conference*, B-2 Volume, pp.395, 2001.

33. Yukio Sone, Atsushi Shimada, Atsushi Miyaji, Yutaka Nakamura, Tetsuya Hanzawa, Kazuhiko Isoda: "Development of viscoelastic dampers using styrene elastomers Part 1. Material experiments and dynamic force experiments of full-scale brace dampers," *Japanese Architecture Abstracts of Academic Lectures*, B-2 Volume, pp.815, 2002.

34. Tetsuya Hanzawa, Yutaka Nakamura, Kazuhiko Isoda, Yukio Sone: "Development of viscoelastic dampers using styrene-based elastomers part 2. Construction of mechanical models," *Abstracts of Academic Lectures at the Japan Architecture Conference*, B-2 Volume, pp.817, 2002.

35. Yutaka Nakamura, Tetsuya Hanzawa: "Development of viscoelastic damper using styrene-based elastomer part 3. Examination of vibration control effect by performance-specified layout design," *Abstracts of Academic Lectures of the Japan Building Conference*, B-2 Volume, pp.819, 2002.

[REFERENCES ON STEEL DAMPERS] FOR EXAMPLE

36. Yosuke Shimawaki, Kenichi Oi, Koichi Takanashi, Hideo Kondo, Kiyoshi Tanaka, Yasuto Sasaki: "Study on vibration properties of steel frame with history damper Part 1 Repeated loading experiment of shear type vibration damping damper using ultra-low yield point steel," *Architectural Institute of Japan Academic Lecture Abstracts*, C-1 Volume, pp.795, 1996.

37. Keiji Sakao, Toshio Maekawa, Hisayoshi Ishibashi, Yasuo Ichinohe: "Development of RC building damper using ultra-low yield point steel (part 1) shear yield type damper," *Abstracts of Academic Lectures of the Japan Building Conference*, B-2 Volume, pp.849, 1997.

38. Toshio Maekawa, Keiji Sakao, Hisayoshi Ishibashi, Yasuo Ichinohe: "Development of RC damper using ultra-low yield point steel (part 2) slit type damper," *Abstracts of Academic Lectures of the Japan Building Convention*, B-2 Volume, pp.851, 1997.

39. Yasuhito Sasaki, Kiyoshi Tanaka: "Static shearing experiment under tensile/ compressive constant axial force of shear panel type damper using ultra-low yield point steel," *Abstracts of Academic Lectures of Japan Building Conference*, C-1 Volume, pp.785, 1999.

40. Shinichi Sawaizumi, Tanemi Yamaguchi, Ryosuke Misato, Masahiro Nagata: "Experimental study on low recycling fatigue characteristics of low yield point steel materials for dampers," *Abstracts of Academic Lectures of the Japan Building Conference*, C-1 Volume, pp.433, 2000.

41. Yukinobu Kurose, Masami Tozawa, Hiroki Sato, Kazushi Shimazaki: "Study of boundary beam damper using low yield point steel part 1 experimental design/static force experiment," *Abstracts of Academic Lectures of Japan Architecture Conference*, C-1 Separate volume, pp.1031, 2002.

42. Hiroki Sato, Yukinobu Kurose, Hitoshi Kumagai, Kazushi Shimazaki: "Study of boundary beam damper using low yield point steel part 2 evaluation of experimental results," *Abstracts of Academic Lectures of the Japan Architecture Conference*, C-1 Volume, pp.1033, 2002.

43. Hiromi Takenaka, Sachi Omiya, Takayuki Teramoto: "Study on the responsiveness of super high-rise buildings using low yield point steel dampers," Separate volume, pp.175, 2004.
44. Sachi Omiya, Takayuki Teramoto: "Study on response properties of high-rise buildings receiving repeated seismic ground motion response properties and fatigue damage of buildings using low yield point steel dampers," *Abstracts of Academic Lectures of the Japan Building Conference*, B-2 Volume, pp.177, 2004.

Chapter 2

Development of an oil damper for buildings considering damping characteristics and various dependencies

2.1 INTRODUCTION

In general, oil dampers efficiently convert vibration energy into heat energy and dissipate it to the outside air, so it has been widely used in industry since the 1940s as a compact and high performance damping device, mainly for automobiles and railways.

Here, the basic principle of a general oil damper will be briefly described. The structure of a general oil damper is a piston structure typified by a water gun or a syringe as shown in Figure 2.1. The piston is pushed by disturbance such as vibration, and when the internal oil flows and the oil passes through the valve, the internal pressure rises, and resistance is generated in the piston. This resistance force acts as a "damping force" and exhibits damping characteristics that depend on the speed. In other words, when the piston is pushed at a low speed, the oil flows slowly from the valve and the resistance acting on the piston is small, and conversely, when the piston is pushed at a high speed, the oil spouts vigorously from the valve and the resistance acting on the piston also increases. It is a device that utilizes the property of speed dependence.

The building oil dampers targeted in this study are those in which oil dampers are installed in series with the brace so that they can be directly attached to the building structure. [1, 2] In other words, it is different from the one in which the functional parts of the oil flow path, which were conventionally attached to the outside, are tackled in the internal axial direction, and the outer shape is cylindrical and has a smart shape.

Here, the development of an oil damper for construction based on this development concept is shown, and the examination result of the validity for future design utilization is shown.

DOI: 10.1201/9781003290261-3

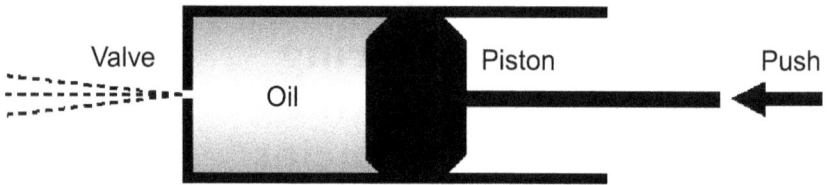

Figure 2.1 Basics principle of general oil dampers.

2.2 DEVELOPMENT CONCEPT AND STRUCTURAL OUTLINE OF BUILDING OIL DAMPER

2.2.1 Development concept

From the standpoint of future design and utilization, the development concept[3, 4] was devised to develop a vibration-damping device that can achieve the following three purposes at the same time, and to improve the building performance.

- Improved livability against vibration in the primary mode of construction caused by daily wind and traffic vibration.
- Improved safety against earthquakes.
- System integrated with design (vibration-damping device to show).

Most of the examples in use in conventional vibration control devices are mass-added vibration control devices such as mass dampers for the level of daily wind and traffic vibration, and history dampers for the level of seismic motion. In many cases, a typical energy absorption type vibration-damping device is adopted, and there are few effective devices except for some viscous dampers that exert a wide range of effects from minute amplitude to large amplitude. In addition, the viscous damper has the demerit that the design becomes complicated because the temperature dependence must be taken into consideration when it is installed in a place where the heat influence is large, such as the window side in the design plan.

Therefore, based on the above-mentioned development concept, we decided to develop a new vibration-damping device that can satisfy both design and structural requirements. The required performance focused on at that time was the following five points.

(1) Maintenance-free equipment that basically does not require replacement.
(2) A device that is not affected by temperature because it may be installed on the window side.
(3) A device with reproducible performance and easy analysis modeling.
(4) A device that can exert effects from small amplitude to large amplitude.

(5) A device that can be installed so that the damping force does not adversely affect the structural frame.

Of the above, it was found that (1) to (3) can be satisfied by adding the following conditions to the existing oil damper. For (1), it is not necessary to replace the damper. However, it is necessary to carry out an emergency visual inspection as a confirmation that there is no abnormality in the mounting condition in case of a change in the situation around the damper, a fire, or an accident that may occur during the period of use of the building. For (2), uses oil used in aircraft and cold regions. For (3), allows modeling with the Maxwell model by mechanically adjusting the internal valve of the oil damper. It is a point. Regarding (4), pressurization is applied to the oil inside the oil damper to crush the air mass existing in the oil from the initial stage to reduce the effect, and the adjustment spring is used for the internal pressure control valve. By incorporating it in double winding, the structure is such that the phenomenon that oil leaks through the gap of the pressure regulating valve and passes through even at a minute amplitude is prevented by the weak adjustment spring of double winding. Regarding (5) at the end, a valve called a release valve, which opens when a load exceeding a predetermined value is reached, is installed to act as a failsafe in which the damping force is suppressed, and as shown in Figure 2.2. The problem was solved by attaching the oil damper part directly to the skeleton like a brace. In addition, to directly attach the oil like a brace, in the conventional structure where the oil reservoir is arranged on the outside of the cylinder, the oil reservoir is horizontal when the oil damper part is slanted. It turned out that it was difficult to keep the shape. For this, we decided to install an oil reservoir inside the cylinder and develop a building oil damper that can

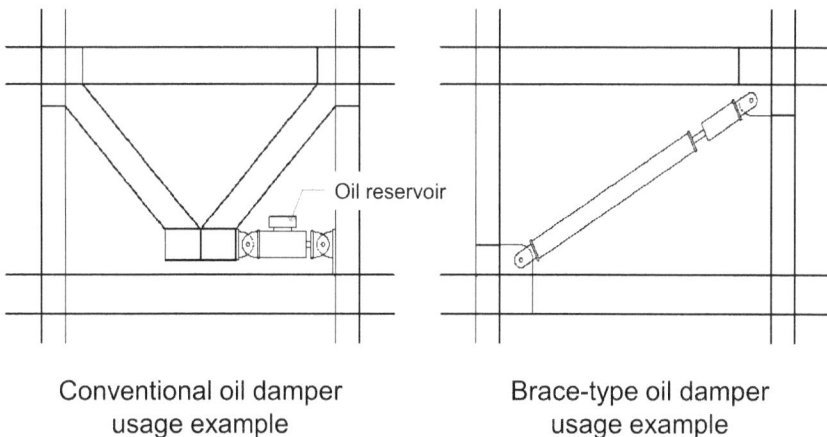

Conventional oil damper
usage example

Brace-type oil damper
usage example

Figure 2.2 Usage examples of oil dampers.

be slanted without any problem. As a result, we also considered that the shape would be smart in terms of design.

2.2.2 Structural outline of oil damper for construction

Figure 2.3 shows the structural outline and shape of the building oil damper[5] targeted in this study. Figure 2.4 shows the hydraulic circuit diagram. This device mainly consists of a cylinder (1), a piston (2), a piston rod (3), an oil reservoir (4), and a seal (5).

In the building oil damper, the piston rod ③ directly connected to the piston ② that slides on the inner surface of the cylinder ① is compressed and expanded in the axial direction when there is an input due to vibration disturbance. At that time, the oil inside passes through the pressure regulating valve ⑥ built in the piston ② and moves to the left and right in the pressure chamber ⑧. At that time, a damping force is generated, but when the load exceeds the specified value, the release valve ⑦ opens and plays a role of suppressing the load increase. This oil damper's feature is to generate a speed-proportional damping force in a minute amplitude region and a relatively low speed region by incorporating double coil springs in the pressure regulating valve ⑥. Furthermore, it is also feature to control the pressurization of the pressure chamber ⑧ and the increase in oil volume with the temperature change of the pressure chamber ⑧ due to the vibration disturbance input by the free piston ⑨ in the piston rod ③. In addition, when joining to the building frame, by adopting a taper pin ⑩ for the spherical bearing ⑪, the joining with the brace mounting plate ⑫ was made into an on-site joint, and a metal touch joint without slack was adopted.

Figure 2.3 (b) shows an enlarged view of the pressure regulating valve ⑥ in which the adjusting spring is incorporated in the double winding.

The pressure regulating valve ⑥ has a structure in which a strong spring (a) is wound outward, and a weak spring (b) is wound inward, and they are incorporated in parallel. The pressure regulating valve is pressed against the orifice by the weak spring (b). At this time, the strong spring (a) is manufactured to have a gap with the pressure regulating valve. A weak

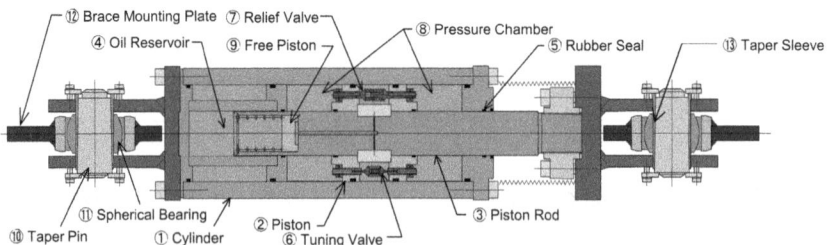

Figure 2.3(a) Inner structure of oil dampers.

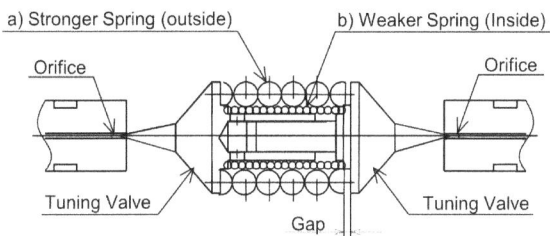

Figure 2.3(b) Detailed view of pressure tuning valves.

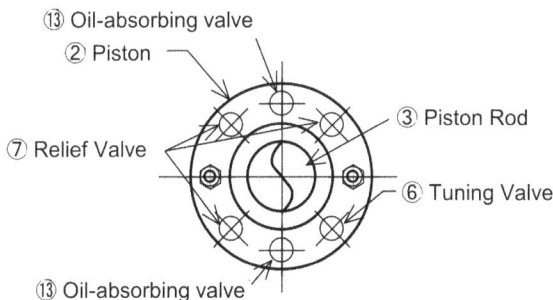

Figure 2.3(c) Detailed view of piston pressure receiving surface.

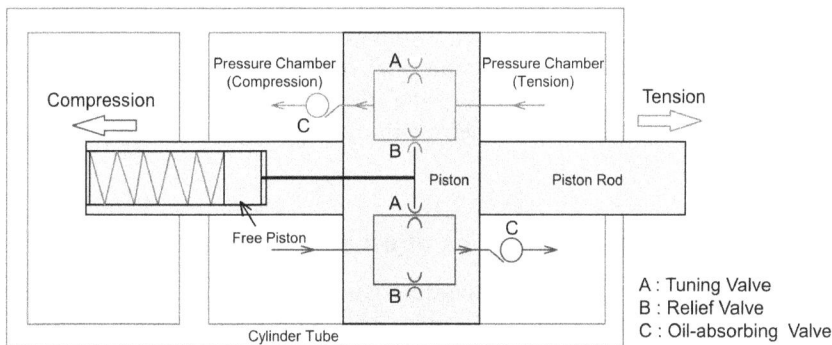

Figure 2.4 Hydraulic circuit diagram.

spring operates in a small amplitude (low speed range), and the mechanism is to achieve the damping characteristics in that range. When the amplitude increases (the speed increases), the strong spring meets the pressure regulating valve, and the mechanism is to achieve the damping characteristic in the speed region after that. That is, in the conventional oil dampers, if the orifice is blocked with a strong spring from the beginning, the operation is like that of a release valve, and the valve does not open until attaining the pressure at which the spring contracts at a constant speed. To avoid this, a

Table 2.1 Comparison of characteristics of silicone oil and petroleum oil

Item		Silicone oil	General petroleum-based oil (reference value)
Relative weight	15/4 °	0.968	0.85
Flash point	COC, °	315	156
Viscosity	cSt @40 °	37.8	10.05
	@100°	15.9	2.997
Viscosity index		342	167
Flow point	°	−55	−40
Total acidity	mgKOH/g	0	1.31
Bubbling	ml @24°	—	40(0)

gap is provided between the strong spring and the pressure regulating valve. Therefore, oil leakage occurs in a minute amplitude (low speed range), and it is difficult to achieve a predetermined damping characteristic. In order to solve this, a weak spring (b) was provided, and the adjustment spring was incorporated into the double winding.

Regarding oil, (1) there is little change over time, (2) there is little change in performance due to temperature, and (3) it is flame-retardant and has a self-extinguishing function unless thermal power is continuously applied from the outside, so it is possible to prevent the spread of fire. Considering this, we decided to use silicon oil, although it is more expensive than the petroleum-based oil used in conventional oil dampers. Table 2.1 shows a comparison of the characteristics of the silicone oil used in the building oil dampers of this study and the petroleum oils used in general oil dampers as a reference.

2.2.3 Design characteristics of oil damper for building

Figure 2.5. shows the design characteristics of the building oil damper used in the experiments in this study. The design characteristics can be expressed by the relationship between damping force: Fd and velocity: ud in the dashpot of the building oil damper. The building oil damper has a bilinear characteristic in which the release valve is opened, and the damping force is suppressed for a load exceeding a predetermined value by providing a release valve in consideration of fail safety. The damping characteristics at that time are set as the maximum damping force: F2 (500 kN), the primary damping coefficient: C1 (125 kN sec./cm), the release damping force: F1 (400 kN), and the relief speed: vr (Secondary attenuation coefficient after passing 3.2 cm/sec.): C2 is 0.068 times the primary. These characteristics are set based on the seismic motion level, with the damping force of the oil damper being generally less than the relief damping force at the rare seismic

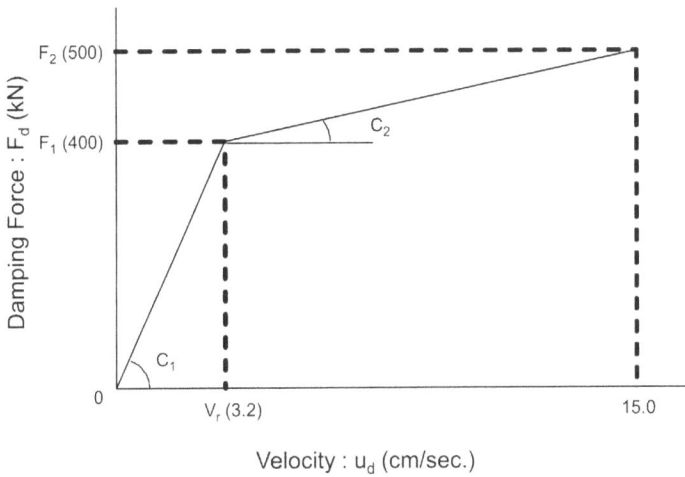

Figure 2.5 Damping characteristics of dashpot (design specification).

motion level and being generally more than the release damping force at the extremely rare seismic motion level.

In addition, regarding the amplitude stroke, it was set to $\pm$ 80 mm in consideration of the construction error and the followability up to the inter-layer deformation angle of about 1/50 on the premise that it is installed as a vertical brace.

2.3 OUTLINE AND RESULTS OF SINGLE EXPERIMENT

2.3.1 Test specimen and experiment method

In the experiment to confirm the damping characteristics of this oil damper, a dynamic vibration test was performed using a vibration test device as shown in Figure 2.6(a). Damping force: F_d is the load cell in the servo cylinder, and displacement: um is the displacement of the total length of the damper and the displacement of the piston rod measured by the laser displacement meter. The experiment was conducted in May 1997 in an environment with an average temperature of 17.1 °C.

The vibration period in this dynamic vibration experiment was set to 1.5 sec. to 4.0 sec. This is mainly unique to skyscrapers with a building height of 60 m or more, where mass-added vibration-damping devices such as mass dampers are often installed as a measure against daily wind and traffic vibration levels. This is because the target is the cycle. The amplitude was also set as shown in Table 2.2 in consideration of the amplitude of the response level at the time of a large earthquake from the slight vibration level.

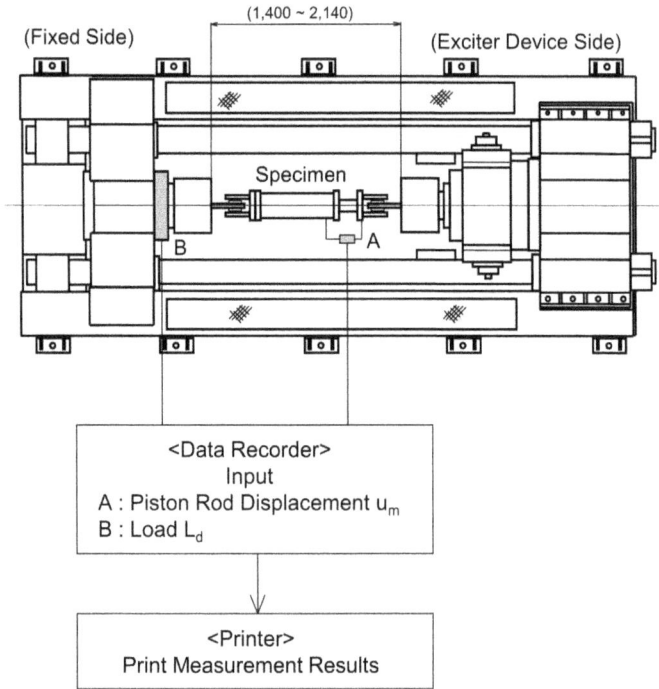

Figure 2.6(a) Exciter device.

Figure 2.6(b) Exciter device.

Table 2.2 Vibration period and maximum amplitude

Cycle [sec.]	1.5	3.0	4.0
Max amp. [mm]	0.25	0.25	0.25
	0.50	0.50	0.50
	0.75	—	—
	1.00	1.00	1.00
	2.50	—	—
	5.00	5.00	5.00
	10.0	10.0	10.0
	15.0	—	—
	20.0	20.0	20.0
	—	50.0	50.0
	—	60.0	—
	—	—	75.0

2.3.2 Experiment results

Several figures are shown below to indicate the basic characteristics of the oil damper for buildings. Figure 2.7 shows the hysteresis loops for the relationship between damping force: *Fd* and displacement: *Um* at 1.5 sec. of the period for the oil damper, and Figure 2.8 shows the relationship between damping force: *Fd* and velocity: *Um* at the same period. The velocity: *Um* was calculated from the inclination of the displacement: *Um* was observed by a displacement meter. In addition, Figure 2.9 shows the waveforms of time-history analyses. The hysteresis loops from the results of the dynamic excitation test showed are elliptical shapes and peculiar to velocity-dependent dampers. Therefore, a simple dashpot model is not sufficient for modelizing.

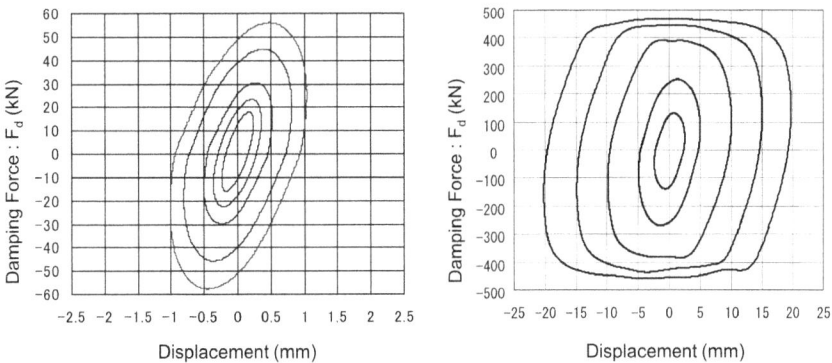

Figure 2.7 Damping force-displacement relationship (period: 1.5 sec.).

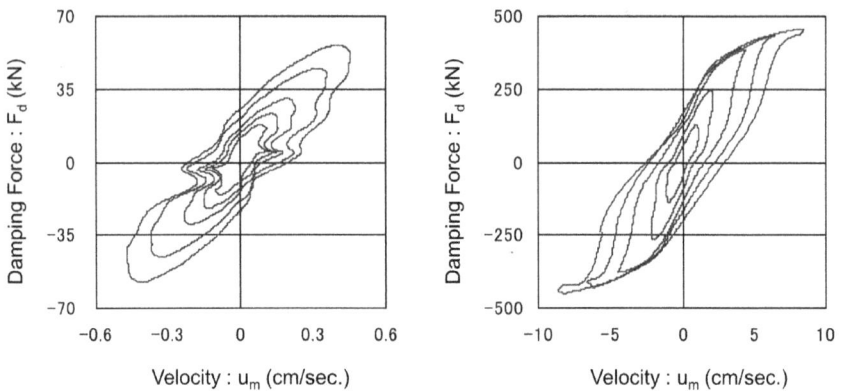

Figure 2.8 Damping force-velocity relationship (period: 1.5 sec.).

However, it predicted that behavior is possible to modelize with the Maxwell model for the following reasons:

• The amplitude dependence observed by the damping force at the maximum displacement.
• The displacement at the maximum damping force formed an almost identical regardless of the amplitude.

Figure 2.10 shows the graphs of the relationship between damping force: Fd and velocity: Ud at 1.5 sec. of the period. Within these graphs, it is plotted for comparison that relationship between the damper damping force: Fd that averaged over the maximum values of either compression and tension sides and the velocity: Um for each hysteresis loop.

As a result, in the vibration period of 1.5 sec. and the range of the damper speed um from the low-speed region with the minimum vibration amplitude of 0.25 mm and 0.1 cm/sec. to the high-speed region with the maximum vibration amplitude of 20.0 mm and 8.4 cm/sec., it was confirmed that the variation error from the design solid line is within ± 10%. Figure 2.7 shows the damping force-displacement relationship (1.5 sec. of vibration period), Figure 2.8 shows the damping force-velocity relationship (1.5 sec. of vibration period), Figure 2.9 shows graphs of time history analysis, and Figure 2.10 shows the damping force-velocity relationship (1.5 sec. of vibration).

Cycle	Amplitude	Damping Force F_d (kN)	Displacement u_m (mm)	Velocity u_m (cm/sec.)
1.5 (sec.)	±0.25 (mm)			
	±0.35 (mm)			
	±0.50 (mm)			
	±0.80 (mm)			
	±1.00 (mm)			
	±2.50 (mm)			
	±5.00 (mm)			
	±10.0 (mm)			
	±15.0 (mm)			
	±20.0 (mm)			

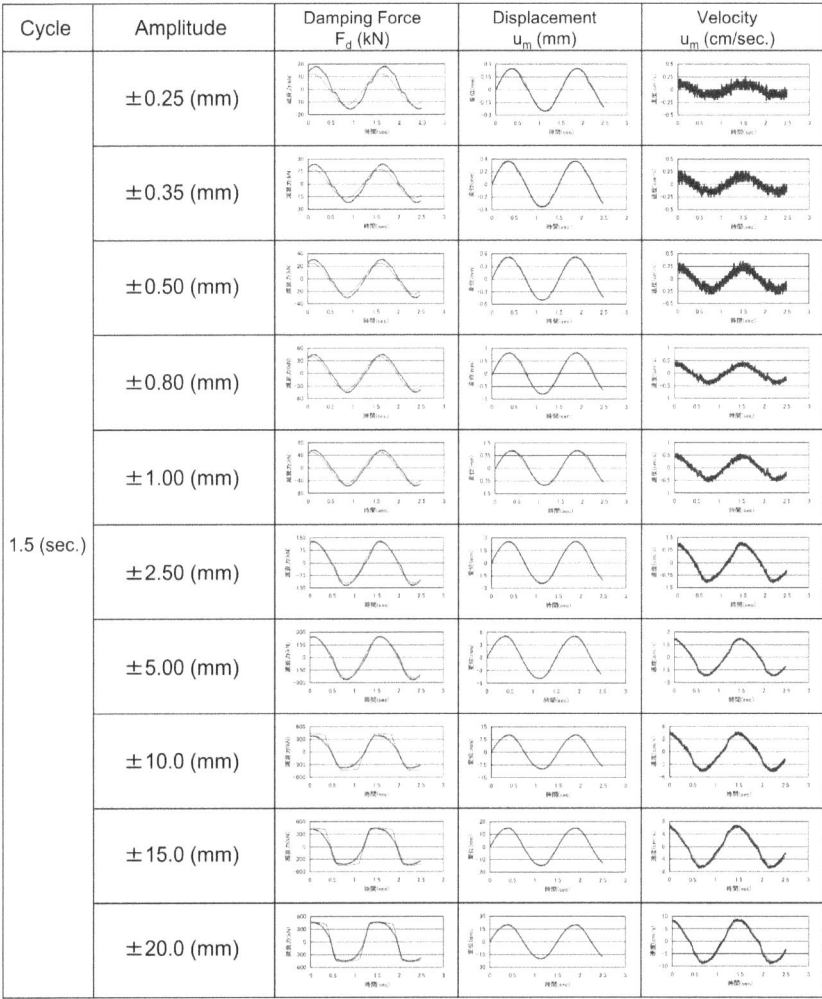

Horizontal Axis : Time (sec.)

Figure 2.9 Time history graphs (experimental results).

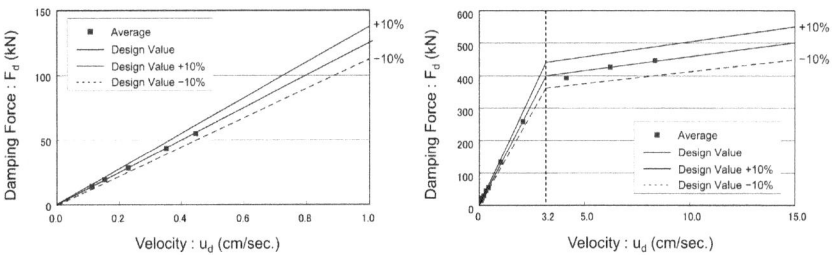

Figure 2.10 Damping force-velocity relationship (Period: 1.5 sec.).

2.4 STUDY ON DAMPING CHARACTERISTICS AND VARIOUS DEPENDENCIES OF SINGLE EXPERIMENT

The following four types of experiments were conducted[6, 7] to confirm various dependent characteristics.

- Temperature dependence confirmation experiment.
- Temperature rise confirmation experiment.
- Durability confirmation experiment.
- Random wave response confirmation experiment.

The results of these experiments are shown below.

2.4.1 Experiment to check temperature dependency

In order to confirm the temperature dependence, a dynamic vibration experiment was conducted with the temperature of this oil damper from –20°C to +80°C every 10°C for a period of 1.5 sec. and an amplitude of 15 mm. In the low temperature range, after cooling in a constant temperature bath at –30°C for 3 days and nights, the oil damper was taken out from the constant temperature bath and the experiment was carried out when the surface temperature of the oil damper reached a predetermined temperature. Figure 2.11 shows the status of the experiment.

Figure 2.12 shows the damping force-displacement relationship between –20°C and room temperature (+20°C) and +80°C. From this result, the damping force: F_d-displacement: um relationship is almost the same in the temperature range of –20°C to +80°C, and it was confirmed that if the temperature is within this range, it is not necessary to specially consider the temperature dependence at the time of designing.

Immediately after
taking out from the thermostatic tank

Vibration test at low temperature

Figure 2.11 Temperature-dependency experiments.

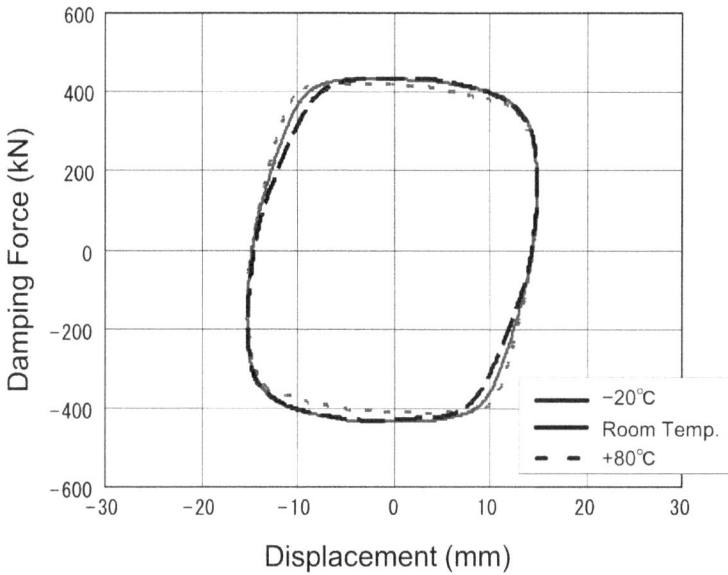

Figure 2.12 Damping force-velocity relationship by temperature (period: 1.5 sec., amp.: 15 mm).

2.4.2 Experiment to check temperature rise

In order to confirm the temperature rise of this oil damper, assuming the duration against shaking in a strong wind of a skyscraper, continuous vibration is performed for a maximum of 8 hours with a period of 4.0 sec. and an amplitude of 1.0 mm, and the oil damper is used. The surface temperature was measured.

Figure 2.13 shows the temperature rise with respect to the elapsed time currently. From this result, the temperature gradually rises from the initial

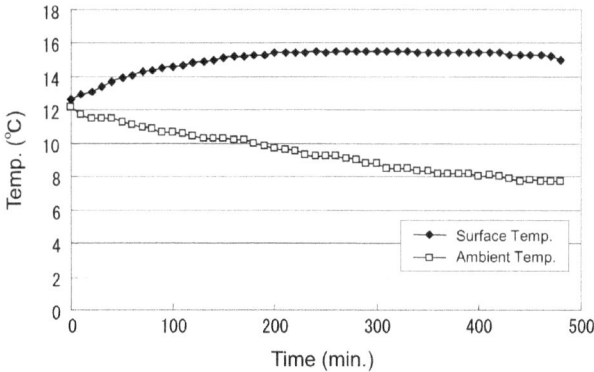

Figure 2.13 Temperature history in excitation experiment (period: 4.0 sec., amp.: 1 mm).

damper surface temperature of 12.6°C to reach 15.8°C until about 200 minutes. After that, the heat generation and heat dissipation of the oil damper are balanced and reach a constant temperature. As a result, it was confirmed that even if this oil damper is installed in the building, no special measures are required due to the temperature rise.

2.4.3 Experiment to check durability

Assuming the sliding distance that this oil damper is expected to experience when used in a skyscraper with a period of 4.0 sec. and 40 floors for 100 years as shown in Table 2.3, the experiment of 1000m sliding distance was carried out. To confirm the above durability, a continuous vibration experiment was conducted 2.5 million times (sliding distance 1000 m) with a period of 0.2 sec. and an amplitude of 0.2 mm.

Regarding the setting of the vibration cycle, if this test is carried out with a cycle of 1.5 sec., it takes about 44 days even if continuous vibration is performed day and night, which is a reality due to the limitation of the occupied time of the testing machine. Therefore, in the case of oil dampers, this experiment is positioned as an accelerated test with the primary purpose of achieving the sliding distance of the seal because the sliding of the seal is critical in terms of durability, and consideration is given to shortening the test time. It was set as day and night continuous vibration for about 6 days with a cycle of 0.2 sec.

In order to confirm the change in damping characteristics before and after this experiment, Figure 2.14 shows a hysteresis loop with a period of 1.5 sec. and an amplitude of 15 mm before and after the durability test. From this result, the hysteresis loops before and after the durability test are almost the same. In addition, it was confirmed that no oil leakage occurred even after the durability test, and it was confirmed that if the building

Table 2.3 Assumed sliding distance (period: 4.0 sec.) (40-story high-rise building)

Earthquakes and wind	Level and size	Duration [sec]	Count	Sliding distance [m]
Major earthquake	50cm/sec	120	1	0.4
Medium earthquake	25cm/sec	60	2	0.2
Small earthquake	10cm/sec	60	100	5.3
Wind	Typhoons about once every 100 years	36000	1	94.5
	Typhoons about twice every 5 years	36000	20	144.0
	About first storm of spring	36000	100	360.0
Sum				598.6

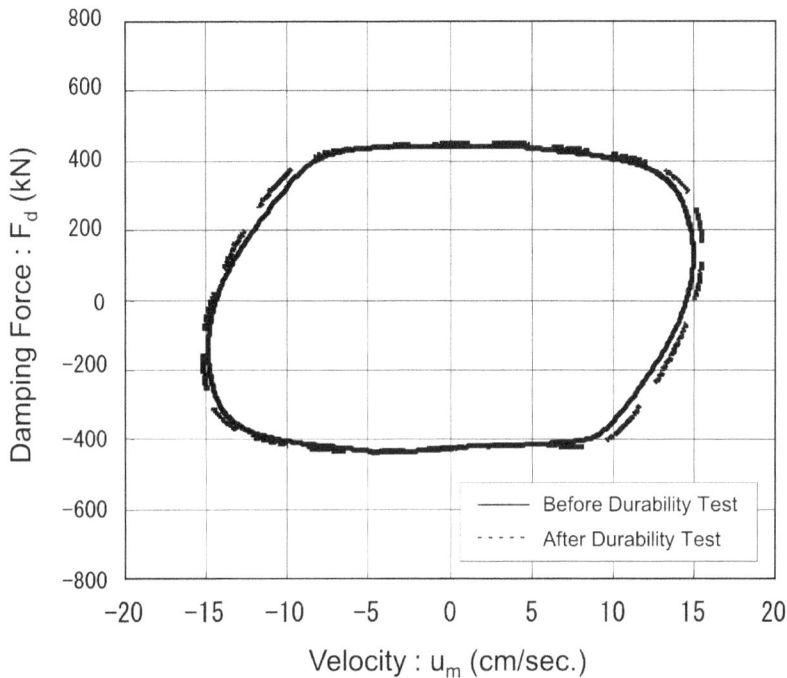

Figure 2.14 Damping force-velocity relationship (period: 1.5 sec.) (before and after durability test).

assumes a sliding distance of about 1000 m, no special consideration is required when using the building.

2.4.4 Experiment to check responses to random waves

As a random wave response confirmation experiment, the acceleration record of the JMA 1995 KOBE NS wave observed at the Ocean Meteorological Observatory of the Hyogoken Nanbu Earthquake was integrated twice to form a displacement waveform, which was standardized to a maximum displacement of 18 mm due to the limit of the testing machine. A dynamic vibration experiment was performed when forced input was performed.

Figure 2.15 shows the damping force: F_d-displacement: um relationship in this experiment.

From this result, it was confirmed that it follows a random input such as a seismic wave without any problem and shows good behavior.

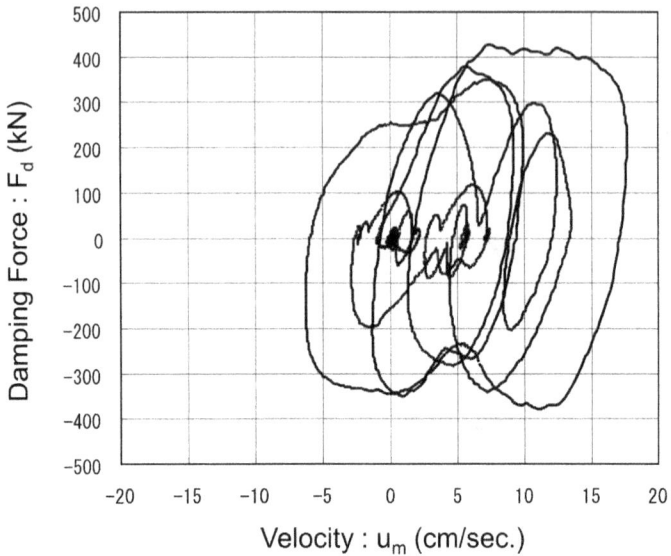

Figure 2.15 Damping force-velocity relationship (random wave).

2.5 SUMMARY

We conducted a systematic dynamic vibration experiment considering the damping characteristics and various dependencies of the building oil damper that can be directly attached diagonally like a brace, and determined the temperature, speed, and frequency. The results of the dependent damping characteristics and durability were organized and confirmed. The main findings obtained are as follows.

(1) It was confirmed that stable performance is exhibited from a low-speed region of about 0.1 cm/sec. by steady sine wave vibration with a period of 1.5 sec.

(2) It was confirmed from the displacement-damping force curve and the velocity-damping force characteristics that the release valve provided separately mechanically also exhibits stable performance as required.

(3) Regarding temperature dependence, it was confirmed that stable performance was exhibited in the range of −20°C to +80°C. The standard temperature for experiments with various dependent characteristics is set at 20°C.

(4) It was confirmed that the temperature rise due to continuous use does not cause a temperature rise that requires special consideration within the range of normal use, and that stable performance is exhibited.

Table 2.4 Evaluation items for oil dampers

Various dependencies	Item	Confirmed range
Oscillation characteristics	Period range	T_1: 1.5, 3.0, 4.0 [sec]
	Amplitude range	±0.25~±75 [mm]
	Damping force range	10~500 [kN]
	Speed range	0.1~8.4 [cm/s]
	Sliding distance range	1000 [mm]
	Temperature dependence	−20~+80 [°]
	Repeat dependencies	1000 [m] Check after cycle shake
Limit characteristics	Speed range	Up to 40 [cm/s]
	Amplitude range	±19.6 [mm] (3.33[Hz])

(5) It was confirmed by a durability test using the sliding distance as an assumed value that stable performance was exhibited without any problem in terms of durability during the period of use of the building. At that time, it was confirmed that there were no abnormalities such as oil leaks.

The building oil damper developed this time is installed directly on the building like a brace to make the building highly damped, be effective from the vibration in the building primary mode caused by the daily environmental vibration level to the response level of a big earthquake. In addition, the purpose is to integrate it into the design and arrange it so as not to adversely affect the surrounding frame as much as possible.

Table 2.4 shows a list of evaluation items for building oil dampers confirmed in this experiment.

REFERENCES

1. Yasuo Rouki, Kunio Furukawa, Yutaka Ishida, Tomio Okabe, Toshiko Okuzono, Osamu Takahashi: "Development of seismic control structure using oil damper (1. Dynamic addition experiment)", *Architectural Institute of Japan Conference Academic Lecture Abstracts (Kyushu)*, B-2 Volume, pp.933–934, 1998.9.
2. Masayuki Ninomiya, Tomio Okabe, Toshiko Okuzono, Osamu Takahashi: "Development of seismic control structure using oil damper (Part 2. Experimental evaluation and analysis model)", *Architectural Institute of Japan Conference Academic Lecture Summary (Kyushu)*, B-2 separate volumes, pp.935–936, 1998.9.
3. Toshiko Okuzono, Osamu Takahashi, Masayuki Ninomiya: "Structural planning institute building; Vibration control structure by directly attaching oil damper", *Steel Structure Technology*, pp.12-1–12-6, 1999.3.

4. Osamu Takahashi, Tomio Okabe, Toshihumi Okuzono, Masayuki Ninomiya: "Response control structure with oil damper bracing system", *The Seventh East Asia on Structural Engineering*, 1999.8.

5. Katsuaki Sunakoda, Naofumi Isobata, Fumiya Iiyama, Izumi Tamura, Toshiko Okuzono, Osamu Takahashi: "Development of hydraulic damper for micro-amplitude (Part 1. dynamic force test)", *Architectural Institute of Japan Conference Academic Lecture Abstracts* (China), B-2 Volume, pp.1061–1062, 1999.9.

6. Ryuichi Inoue, Toshiko Okuzono, Osamu Takahashi, Fumiya Iiyama, Naofumi Iwahata, Kazuhiko Shibata: "Development of hydraulic dampers for minute amplitude (Part 2. Comparison of dynamic force test results and analysis results)", *Abstracts of Academic Lectures at the Japan Architecture Conference (Tohoku)*, B-2 Volume, pp.863–864, 2000.9.

7. Kazuhiko Shibata, Fumiya Iiyama, Naofumi Isobata, Toshiko Okuzono, Osamu Takahashi, Ryuichi Inoue: "Development of hydraulic damper for micro-amplitude (Part 3. Dynamically applied test (2))", *Architectural Institute of Japan Abstracts of Academic Lectures (Tohoku)*, B-2 Volume, pp.865–866, 2000.9.

Chapter 3

Analytical model of a single building oil damper

3.1 INTRODUCTION

In recent years, oil dampers have been used in many buildings in a broad sense as vibration-damping members for the purpose of highly damping the buildings and absorbing vibration energy.

The building oil damper targeted in this research is a newly developed building oil damper incorporated in series in the brace so that it can be directly attached to the building structure. In the brace-type oil damper targeted this time, a relief valve is provided so that the damping force transmitted from this vibration-damping system to the peripheral frame does not become excessive, and the damping force is suppressed.

It is desirable that the analysis model of these vibration-damping systems can reproduce the actual dynamic behavior as faithfully as possible. In addition, the parameters of the analysis model of a general oil damper were often provided by the manufacturer in the past, but when applying it, it is desirable to confirm the correspondence with the model by an actual machine test.

Here, we show an analytical model method that evaluates the experimental results carried out during the development of a building oil damper based on this development concept and a numerical calculation algorithm that calculates the damping force and verify the accuracy of the calculation method. It shows the result of the examination of the validity for future design utilization.

3.2 OVERVIEW OF THE ANALYTICAL MODEL AND NUMERICAL CALCULATION ALGORITHM

3.2.1 Basic properties for modeling building oil damper

From the damping force-displacement (period 1.5 sec.) hysteresis loop obtained from the dynamic vibration experiment, the shape shows an

DOI: 10.1201/9781003290261-4

elliptical shape peculiar to the velocity-dependent damper that is insufficient to be modeled by a simple dashpot model, and the amplitude dependence was confirmed for the equivalent stiffness K_0 at the maximum displacement shown in Figure 3.2. On the other hand, since the rigidity K_d at the maximum damping force does not change significantly, this damper is necessary to be modeled by Maxwell element in which a linear spring and a nonlinear dashpot are connected in series as shown in Figure 3.1.[1, 2] The history curve of the Maxwell element draws a tilted ellipse as shown in Figure 3.2. In the figure, K_d indicates the spring rigidity, and the equivalent rigidity K_0 can be expressed by the following equation using the damping coefficient C of the dashpot and the circular frequency ω.

$$K_0 = \frac{\omega^2 K_d C_d^2}{K_d^2 + \omega^2 C_d^2} \tag{3.1}$$

3.2.2 Calculation method of Maxwell element force

Much research has been done on the method of calculating the elemental force of the nonlinear Maxwell model.[3, 4] Here, as shown in Figure 3.1, we propose a method for calculating the elemental force of a nonlinear Maxwell model in which a linear spring with spring rigidity K_d and a nonlinear dashpot with bilinear damping characteristics as shown in Figure 3.3 are arranged in series. The features of this method are that less input data is required for numerical calculation, and when the time step Δt of the analysis time is sufficiently small, the speed at time t is approximated by the average speed from $t - \Delta t$ to t. Then, by assuming that the velocity at time t can be approximated, it is not necessary to perform complicated calculations.

First, the balance of forces at the connection point O is obtained. In the following equations, u represents the relative displacement and v represents the relative velocity. Using the elastic force F_k due to the spring and the viscous force F_d due to the dashpot, the Maxwell element force F_{ij} is expressed by the following equation.

$$F_k = F_d = F_{ij} \tag{3.2}$$

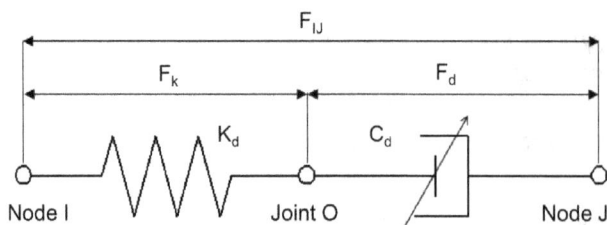

Figure 3.1 Maxwell model.

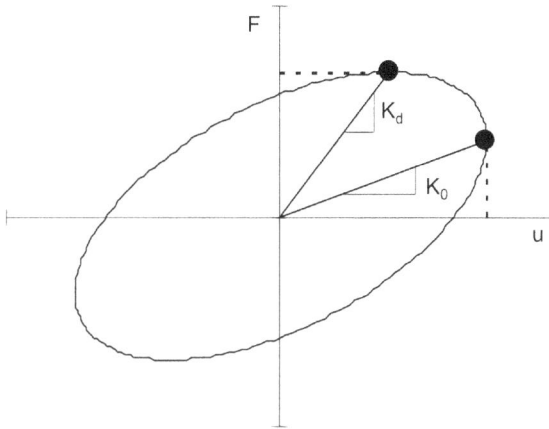

Figure 3.2 History curve of Maxwell model.

The spring force F_k can be obtained by the following equation using the relative displacement u_{ij} between the Maxwell elements and the relative displacement u_{oj} between the dashpots.

$$F_k = K_d \cdot \left(u_{ij} - u_{oj} \right) \tag{3.3}$$

The damper force after relief is expressed by the following equation using the intersection F_c between the secondary gradient and the axis of damping force.

$$F_d = C_2 \cdot v_{oj} + F_c \tag{3.4}$$

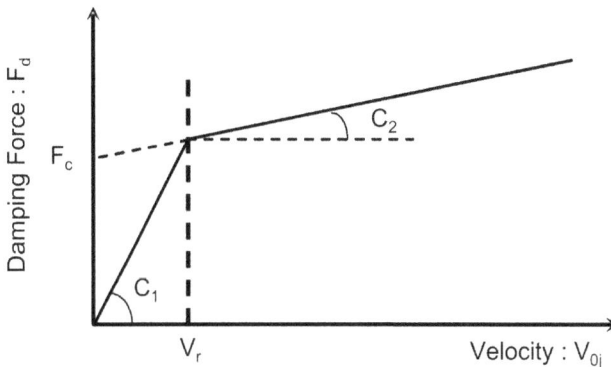

Figure 3.3 Damping force-velocity relationship of dashpots.

Therefore, the equilibrium condition of the connection point O is as follows by substituting Eqs. (3.3) and (3.4) into Eq. (3.2).

$$K_d \bullet \left(u_{ij} - u_{oj} \right) = C_2 \bullet v_{oj} + F_c \tag{3.5}$$

If the time step Δt is sufficiently small, the velocity at time t can be assumed by the following equation if the velocity at time t is approximated by the average velocity from $t - \Delta t$ to t.

$$^t v_{oj} = \frac{\Delta^t u_{oj}}{\Delta t} \tag{3.6}$$

where $^t v_{oj}$ represents the relative velocity between the dashpots at time t, and $\Delta^t u_{oj}$ represents the increment of the relative displacement between the dashpots at time t. Substituting Eq. (3.6) into Eq. (3.5) gives the following equation.

$$K_d \bullet \left({^t u_{ij}} - {^t u_{oj}} \right) = C_2 \bullet \left(\frac{\Delta^t u_{oj}}{\Delta t} \right) - F_c \tag{3.7}$$

If the relative displacement at time t is expressed by the sum of time $t-\Delta t$ and increment, the following equation is obtained.

$$^t u_{oj} = {^{t-\Delta t} u_{oj}} + \Delta^t u_{oj} \tag{3.8}$$

Using Eq. (3.8), Eq. (3.7) can be summarized for $\Delta^t u_{oj}$ as follows.

$$\Delta^t u_{oj} = \frac{K_d \bullet \left({^t u_{ij}} - {^{t-\Delta t} u_{oj}} \right) - F_c}{\dfrac{C_2}{\Delta t} + K_d} \tag{3.9}$$

By substituting the $^t u_{oj}$ obtained by substituting Eq. (3.9) into Eq. (3.8) into Eq. (3.3), the spring force can be obtained, and the Maxwell element force can be obtained.

The Maxwell element force before relief can be obtained by replacing C_2 in the equation with C_1 and setting $F_c = 0$.

3.2.3 Input data required for numerical calculation

The input data required when performing numerical calculation using the above calculation method is as follows.

(a) Parameters of building oil dampers
  - Spring rigidity K_d
  - Attenuation coefficient C_1, C_2
  - Relief speed v_r

(b) Analysis parameters
  • Time step Δt

3.2.4 Numerical calculation algorithm

A numerical calculation algorithm for calculating the damping force when a displacement is applied to a building oil damper will be described with reference to the flowchart shown in Figure 3.4.

First, using the relative displacement $^{t}u_{ij}$ between the given dampers and the relative displacement $t\text{-}\Delta^{t}u_{oj}$ between the dashpots in the previous step, the displacement increment $\Delta^{t}u_{oj}$ between the dashpots within the linear range (before relief) is calculated. This can be obtained by the following equation in which C_2 is replaced with C_1 in Eq. (3.9) and Q_c is set to zero.

$$\Delta^{t}u_{oj} = \frac{K_d \bullet \left({}^{t}u_{ij} - {}^{t-\Delta t}u_{oj} \right)}{\dfrac{C_1}{\Delta t} + K_d} \tag{3.10}$$

Next, the relative velocity v_{oj} between the dashpots is calculated by substituting Eq. (3.10) into Eq. (3.6), and the relief is judged. If this speed exceeds the relief speed on the positive side, the relative displacement increment between the dashpots is recalculated by Eq. (3.9), and if it is lower than the relief speed on the negative side, Q_c is replaced with $-Q_c$ in Eq. (3.9). Recalculate using the following equation.

$$\Delta^{t}u_{oj} = \frac{K_d \bullet \left({}^{t}u_{ij} - {}^{t-\Delta t}u_{oj} \right) + Q_c}{\dfrac{C_2}{\Delta t} + K_d} \tag{3.11}$$

Substituting the $\Delta^{t}u_{oj}$ obtained in this way into Eq. (3.8), the relative displacement u_{oj} between the dashpots and the relative displacement u_{oi} between the springs are calculated. The damping force of the damper is calculated by multiplying the relative displacement between the springs by the spring constant.

If Δt is not sufficiently small in this method, an error will occur in the numerical calculation result of this assumption, and the accuracy will deteriorate. It is necessary to set Δt sufficiently small when calculating numerical values. Figure 3.5 shows the error in the numerical calculation result when Δt is the parameter.

At 0.01 sec. or less in time increments, the theoretical solution is almost the same, and this tendency is the same for input waves with different amplitudes and frequencies. From this, it can be said that this numerical calculation method has sufficient accuracy in the range of 0.01 sec. or less in time increments.

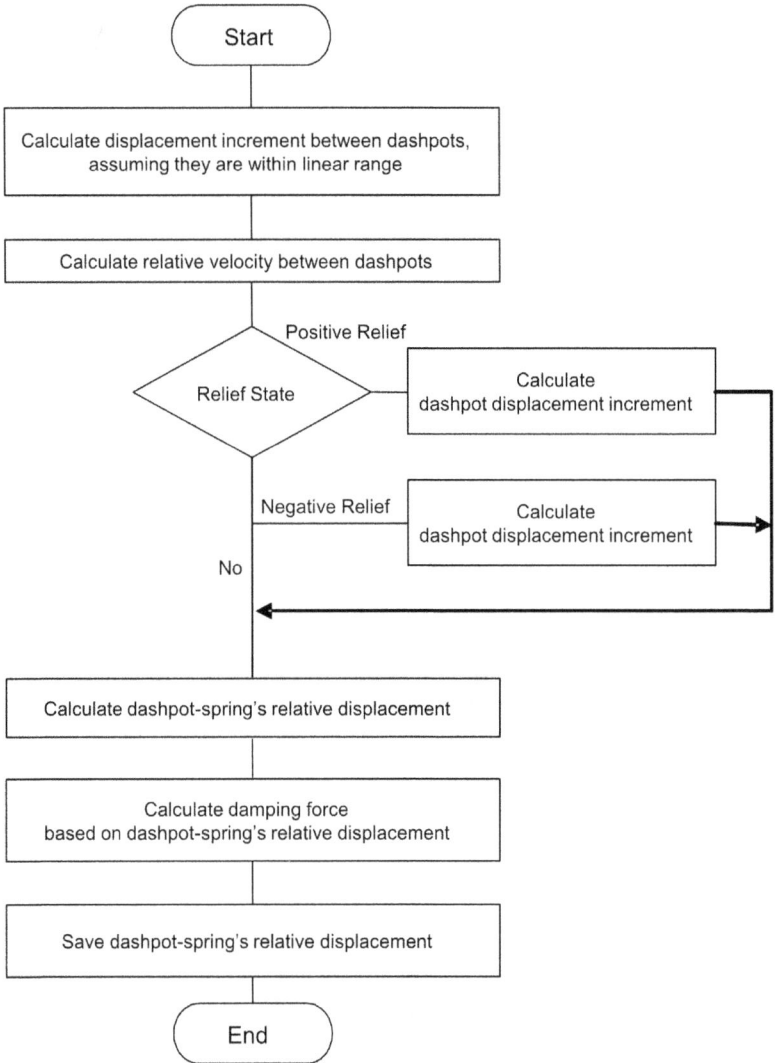

Figure 3.4 Flowchart.

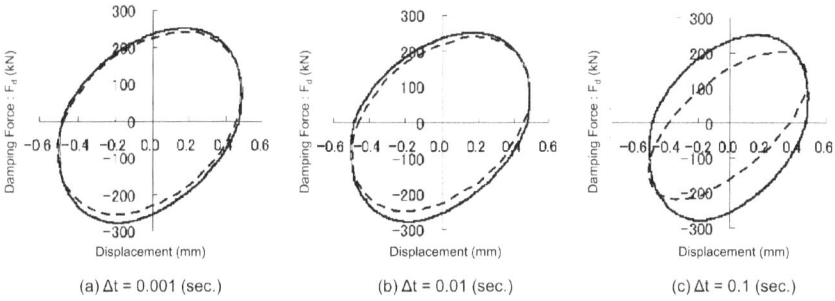

(a) Δt = 0.001 (sec.)   (b) Δt = 0.01 (sec.)   (c) Δt = 0.1 (sec.)

Figure 3.5 Influence of setting of time increment Δt (Experiments and analysis) (Period: 1.5 sec.).

3.3 COMPARISON OF ANALYTICAL MODEL AND EXPERIMENTAL RESULTS OF SINUSOIDAL EXCITATION

To verify the validity of this numerical calculation method, the response analysis at the time of sine wave input was compared with the experimental value. As the parameters of the building oil damper, the following values set at the time of design were used as shown in Figure 3.6. In addition, K_d is determined by the compression rigidity of the oil, but the following values were used from past cases and preliminary experiments at the time of trial production.

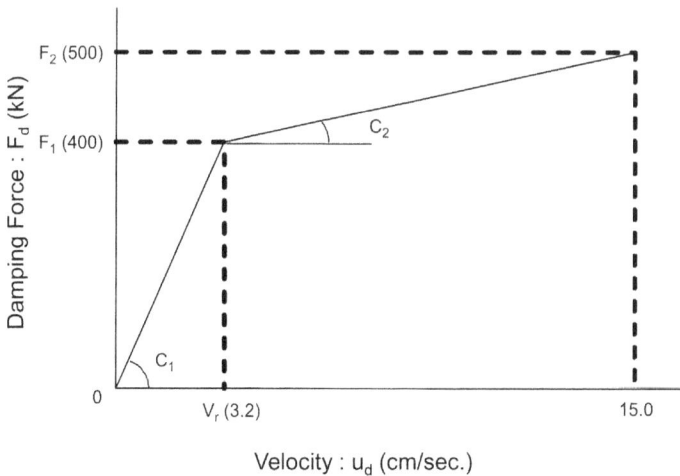

Figure 3.6 Damping force-velocity relationship of dashpots (design specification).

- Spring rigidity K_d = 1400 kN/cm
- First-order attenuation coefficient C_1 = 125 kN·sec./cm
- Secondary attenuation coefficient C_2 = 0.068 C_1
- Release speed V_r = 3.2 cm/sec.

Figures 3.7 and 3.8 show the overlay of the experimental and analysis results when inputting a sine wave. The time step Δt in the analysis was 0.001 sec.

If the amplitude is small, the analysis result and error can be confirmed by the combination of the friction of the seal of the building oil damper and the influence of the weak spring of double winding, but the experimental result has more damping force and hysteresis than the analysis result. Since the entanglement area of the spring is large and the absolute value of the error is small, it was judged to be within the error range on the safe side in practical use. When the amplitude becomes large, it can be said that the experimental result and the analysis result are in good agreement regardless of the frequency and the amplitude.

3.4 COMPARISON OF ANALYTICAL MODEL AND EXPERIMENTAL RESULTS OF RANDOM EXCITATION

Figures 3.9 and 3.10 show the overlay of the experimental and analysis results when random waves are input. In the experiment, the JMA KOBE 1995 NS wave acceleration recording was integrated twice to form a displacement waveform, which was standardized to a maximum displacement of 18 mm due to the limit of the testing machine and input.

From these figures, this numerical calculation algorithm can obtain sufficiently accurate results even for random waves.

3.5 SUMMARY

We conducted a systematic dynamic vibration experiment considering the damping characteristics of a building oil damper that can be directly attached diagonally like a brace and conducted a systematic dynamic vibration experiment to determine the nonlinear damping characteristics by the release characteristics. We compared and confirmed the consideration of the nonlinear dashpot and the analysis model by the nonlinear Maxwell model in which the spring was connected in series. The main findings obtained are as follows.

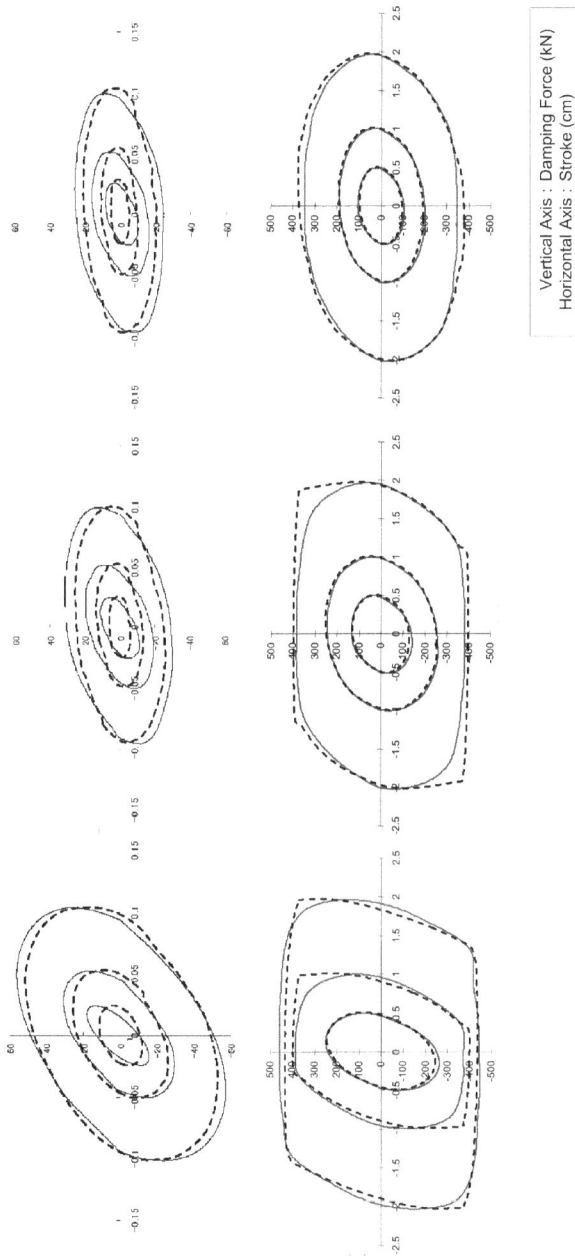

Vertical Axis : Damping Force (kN)
Horizontal Axis : Stroke (cm)

Figure 3.7 Damping force-displacement relationship (experiments and analysis).

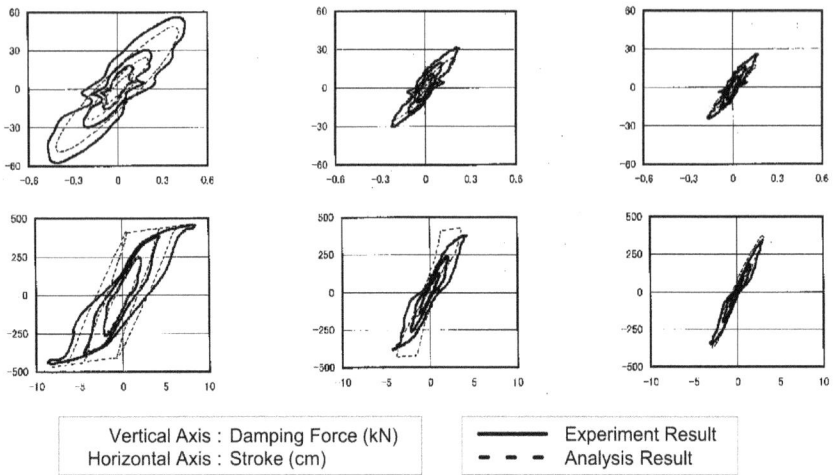

| Vertical Axis : Damping Force (kN) | ——— Experiment Result |
| Horizontal Axis : Stroke (cm) | - - - Analysis Result |

Figure 3.8 Damping force-velocity relationship (experiments and analysis).

Damping Force : F_d (kN)

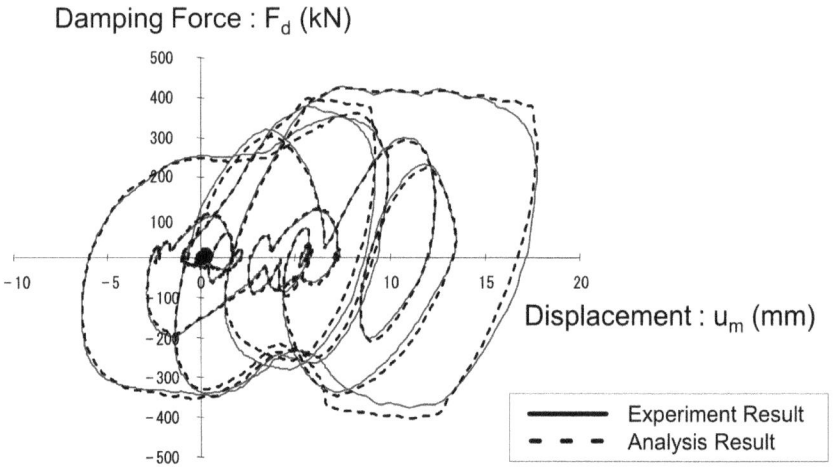

Displacement : u_m (mm)

——— Experiment Result
- - - Analysis Result

Figure 3.9 Damping force-displacement relationship (random wave).

(1) In terms of analysis modeling, the amplitude is measured by sinu-soidal vibration using a nonlinear Maxwell model in which a non-linear dashpot and an oil damper-specific spring are connected in series in consideration of bilinear damping characteristics due to relief characteristics. A good match was confirmed in a large range. It was confirmed that the match was accurate even in the random wave excitation.

(2) In the range where the amplitude due to sinusoidal vibration is small, the experimental results tend to have a larger damping force and

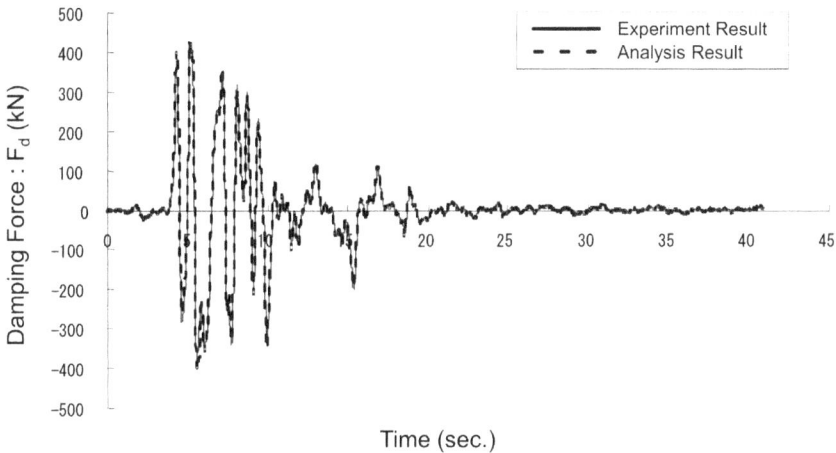

Figure 3.10 Damping force history (random wave).

envelope area of the hysteresis loop than the analysis results. It is a problem with the friction effect of the seal, but how to reflect this effect in the analysis modeling will be examined in Chapter 6.

The method of calculating the elemental force of the nonlinear Maxwell model proposed this time calculates the damping force-displacement relationship of the nonlinear Maxwell model as a single unit from the displacement, so the damping force can be effectively combined with the equation of motion of the structure. It can be considered by adding the term to the equation of motion.

REFERENCES

1. Osamu Takahashi, Yohei Sekiguchi: "Analysis algorithm and Subrutin of oil damper using Maxwell model", *Passive Vibration Control Structure* Symposium 2001 (Tokyo Institute of Technology, Architectural Physics Research Center), pp.101–108, 2001.12.
2. Osamu Takahashi, Youhei Sekiguchi: "Time-history analysis model for nonlinear", SEWC2002, 2002.10.
3. Masatoshi Ishida, et al.: "Study on dynamic response analysis method of structures including nonlinear Maxwell model", *Architectural Institute of Japan Conference Academic Lecture Abstracts* (Kanto), Volume B, pp.655–656, 1993.9.
4. Tomohiko Hatada Takuji Kobori, Masatoshi Ishida, Naoki Niwa.: "Dynamic analysis of structures with Maxwell model", *Earthquake Engineering & Structural Dynamics*, 29, pp.159–176, 2000.

Chapter 4

Damping characteristics and analytical model of brace-type oil damper

4.1 INTRODUCTION

The brace-type oil damper targeted in this study is composed by connecting in series brace member and the building oil damper that discussed in the previous chapters. Furthermore, this damper can be directly attached to the building structure using pin part that joined easily in the construction field. The structure like these can satisfy following two points.

(1) Make sure that the damping force does not adversely affect the structural frame.
(2) Make it easy to remove during construction.

Regarding the appearance of the building oil damper, the oil reservoir, which was conventionally attached to the outside, was made into a cylindrical and smart shape by working in the axial direction inside.

Figure 4.1 shows an example of the use of the brace-type oil damper[1] targeted in this study.

Here, a brace-type oil damper according to this development concept was manufactured with a length assuming that it would be mounted on an actual building, and its damping characteristics were confirmed by experiments. As an experiment using this device, an axial vibration experiment and a frame vibration experiment by incorporating it diagonally into the frame were carried out, and the behavior when receiving a disturbance was confirmed. In addition, based on these experimental results, as an example of contents useful for future design, the basic characteristics of the brace-type oil damper, the shape of the steel pipe brace part, and its influence are shown, and the formulation of the analysis model is proposed.

DOI: 10.1201/9781003290261-5

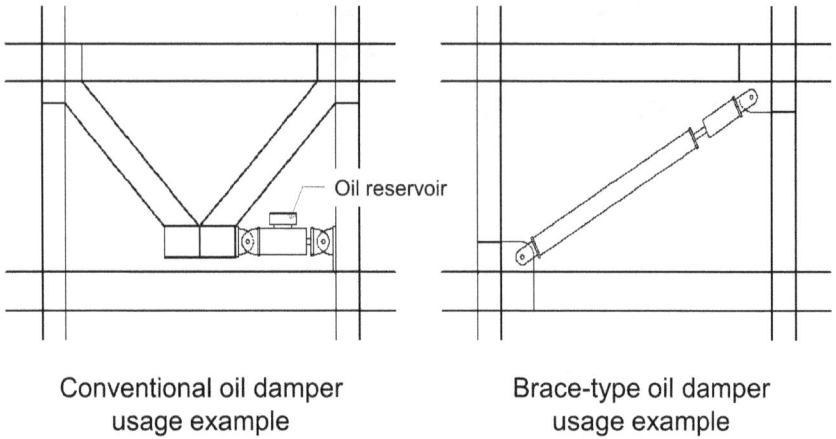

Conventional oil damper
usage example

Brace-type oil damper
usage example

Figure 4.1 Usage examples of oil dampers.

4.2 OVERVIEW OF BRACE-TYPE OIL DAMPER CHARACTERISTICS

4.2.1 Overview of the structure of a brace-type oil damper

Figure 4.2 shows an example of the configuration and arrangement of the brace-type oil damper. Basically, it is a structure in which the oil damper part and the steel pipe brace part are connected in series.

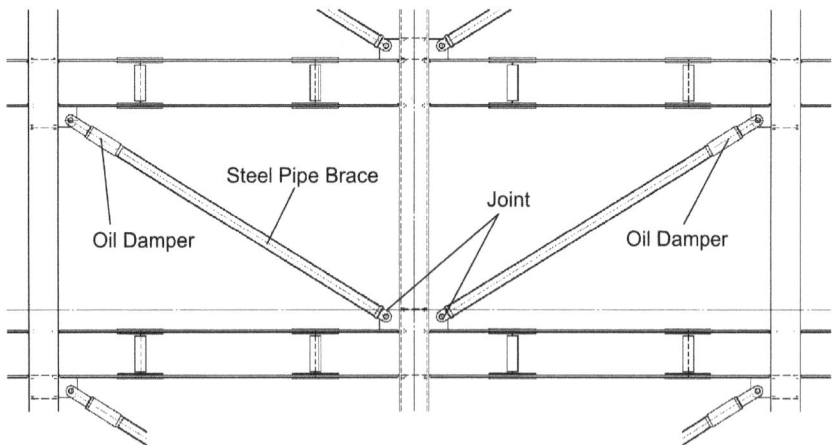

Figure 4.2 Brace-type oil damper installation.

Regarding the outer diameter of the damper, we assumed that the width of the steel beam, which is often used in building planning, is 300 mm, and thought that a size smaller than that size is desirable. Therefore, the oil damper part is a 500 kN type building oil damper (outer diameter = Φ200), and the steel pipe brace part is a building steel pipe (outer diameter = Φ216.3), which are connected in series. This was the subject of consideration.

The end part is a spherical bearing with the expectation that each part will function normally even when the workability of on-site mounting and the force in the out-of-plane direction are applied, as shown in Figure 4.3, and the end part and the damper is connected by a mechanical pin. A mechanical pin with a spherical bearing, which is a combination of this taper sleeve and taper pin, is used to make the gap as small as possible during on-site construction.

In consideration of the processing accuracy of the pin, it was decided to request the machine maker to process the gusset plate attached to the skeleton frame as a set of equipment. After the machined gusset plate and steel frame are welded and joined by a steel frame processing company to be built at the construction site, the brace-type oil damper and gusset plate separately brought to the construction site are installed at the site.

In addition, regarding the amplitude stroke, it was set to ± 80 mm in consideration of the building error and the followability up to the building interlayer deformation angle of 1/50 on the premise that it is installed as a vertical brace.

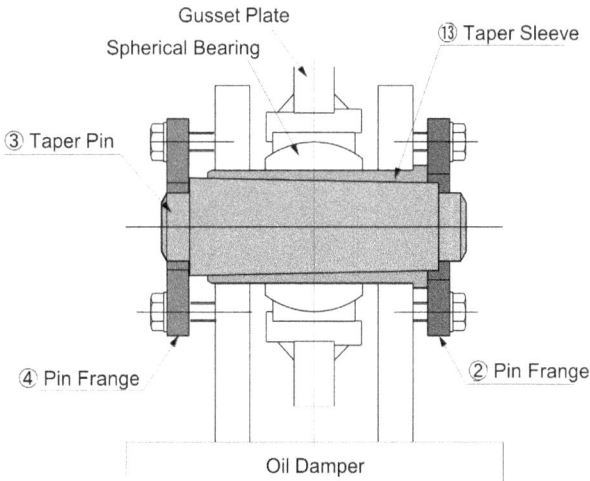

Figure 4.3 Joint.

4.2.2 Basic characteristics of brace-type oil damper

As shown in Figure 4.2, the brace-type oil damper is designed to be diagonally attached to a frame with a span of 7,000 mm and a floor height of 4,000 mm, assuming a general steel-framed office building.

Figure 4.4 shows the basic characteristics of the brace-type oil damper part. As shown in (a), the basic characteristics can be expressed by the dashpot damping force: F_d-velocity: u_d relationship.

In the construction of the oil damper, the relief valve is provided in consideration of the fail safety, so that the relief valve opens and the damping force is suppressed for a load exceeding a predetermined value. The damping characteristics at that time are set to the maximum damping force: Q2 (500 kN), the primary damping coefficient: C_1 (125 kN · sec./cm), and the release damping force: Q1 (400 kN). The attenuation coefficient is 0.068 times the first order. The oil compression rigidity (spring rigidity) is determined by the shape of the oil damper and the capacity of the internal oil: K_d was set to a value (1400 kN/cm) that can be set in manufacturing. Figure 4.4(b) shows an example of the damping force: F_d-displacement: um relationship.

4.2.3 Cross-sectional shape of steel pipe brace

The steel pipe brace part adopts Φ216.3 × 8.2 (STKN400) as a cross section where the buckling resistance is 1.5 times or more of the oil damper-maximum damping force for the mounting length (pin-pin) distance of

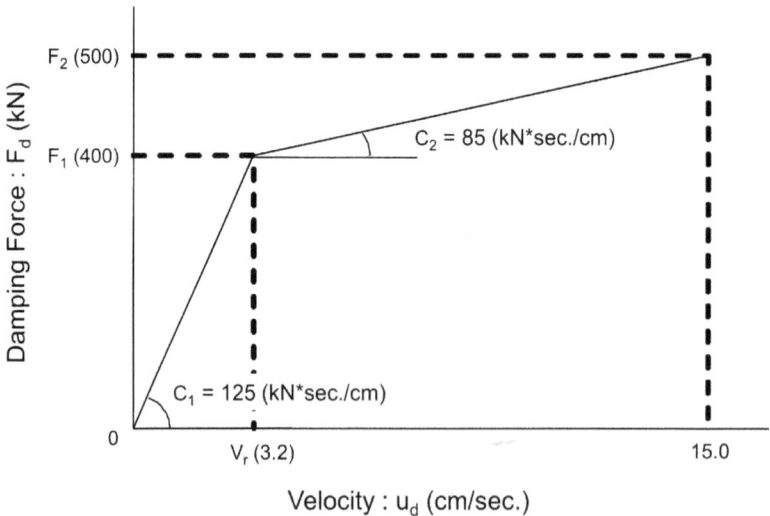

Figure 4.4(a) Damping force-velocity relationship.

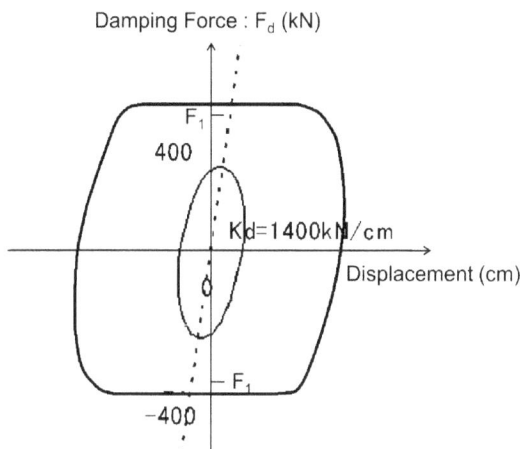

Figure 4.4(b) Damping force-displacement relationship.

6,000 mm. Regarding the setting that the proof stress is 1.5 times or more (safety factor 1.5 or more), the safety factor for the proof stress of general oil damper internal parts as mechanical parts is set to 1.5 or more.

4.2.4 Analytical model of oil damper for construction[2, 3]

The damping force-displacement relationship of the building oil damper with damping characteristics shown in Figure 4.4(b) shows an elliptical shape peculiar to the speed-dependent damper, and modeling is insufficient with a simple dashpot model. In addition, the amplitude dependence can be confirmed for the equivalent rigidity at maximum displacement: K_0 shown in Figure 4.6; the rigidity at maximum damping force: K_d does not show a large fluctuation, so it is handled by numerical calculation. It is well known that a linear spring and a nonlinear dashpot can be modeled into a nonlinear Maxwell element connected in series as shown in Figure 4.5. In other words, the building oil damper can be viewed as a device that can be tuned and manufactured at the time of manufacture so that analysis can

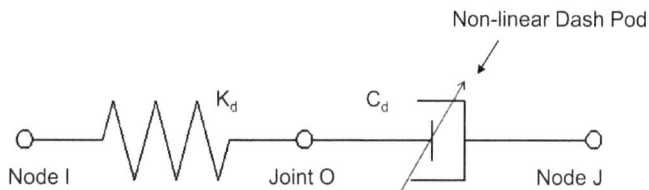

Figure 4.5 Nonlinear Maxwell model.

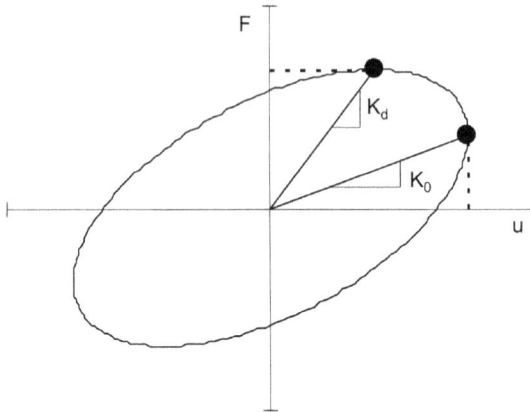

Figure 4.6 History curve of nonlinear Maxwell model.

be easily performed with a nonlinear Maxwell model. The history curve of the Maxwell element draws a tilted ellipse as shown in Figure 4.6. In the figure: K_d indicates the spring rigidity, and the equivalent rigidity: K_0 can be expressed by the following equation using the damping coefficient of the dashpot: C_d and the circular frequency: ω.

4.3 OVERVIEW AND RESULTS OF FULL-SCALE DYNAMIC EXCITATION EXPERIMENTS

4.3.1 Purpose of the experiment

The brace-type oil damper has a longer overall length when mounted on a building, so compared to the experiment of the building oil damper alone, these following problems may be occurred.

(1) Occurrence of buckling.
(2) Resonance of internal parts of oil damper.

Therefore, a full-scale dynamic vibration experiment was conducted with the aim of confirming that these problems did not occur.[4, 5]
    The experiment was conducted in October 1999 in an environment with an average temperature of 18.2°C.

4.3.2 Test specimen and experimental method

In order to confirm the damping characteristics of the brace-type oil damper, a dynamic vibration test was conducted using a vibration test device as shown in Figure 4.7. The damping force was measured with a load cell

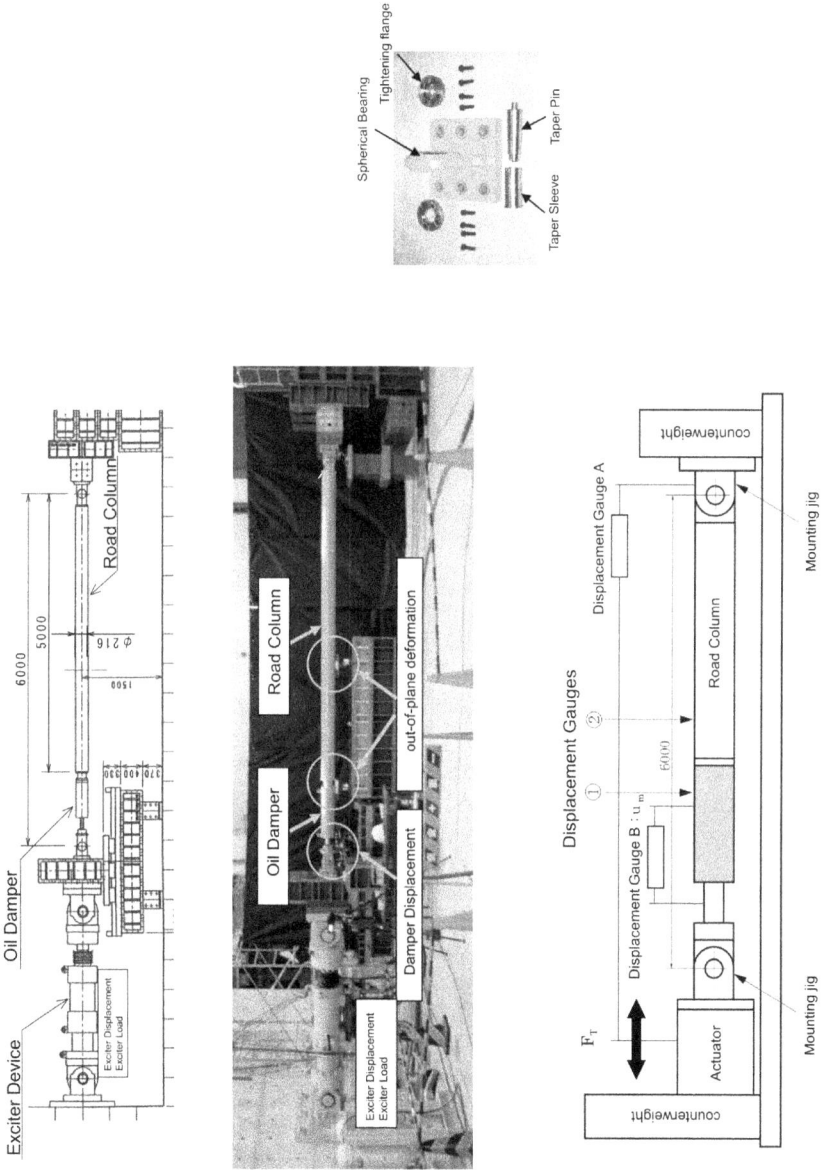

Figure 4.7 Excitation experiment plan.

Table 4.1 Excitation period and amplitude list

Cycle [sec.]	1.5	4.0
Max amp. [mm]	5.0	—
	10.0	10.0
	20.0	20.0
	30.0	30.0
	—	50.0
	—	60.0
	—	70.0

and the displacement was measured with a laser displacement meter. The frequency-dependent parameters in this experiment were set as shown in Table 4.1.

When setting the frequency-dependent parameters, we decided to set them with the image of a building that is likely to be used in a relatively large number of designs, in consideration of its use in future designs. For example, regarding the vibration cycle, imagine a steel-framed 10-story middle-rise office building with a vibration cycle of 1.5 sec., and a steel-framed 40-story skyscraper office building. The case where the vibration period is 4.0 sec. is set.

Regarding the amplitude parameter, the maximum amplitude was set within the range where the interlayer deformation angle was about 1/400 to 1/50, taking into consideration the fact that the brace can be attached diagonally and the allowable capacity of the vibration exciter.

4.3.3 Analytical model for full-scale dynamic excitation experiment

Up to the previous chapter, it has been reported that behaves of a single building oil damper with release characteristics can be modeled by a non-linear Maxwell model.

Since this chapter targets the brace-type oil damper in which the steel pipe part is joined to the building oil damper single part, it is necessary to set the springs other than the building oil damper part to be linear as shown in Figure 4.8. It was calculated as a series spring synthesized as a condition, and the damping force was calculated from the displacement by substituting the parameters of the building oil damper according to the method shown in the flowchart in Figure 3.4 in Chapter 3. The cross-sectional shape of the steel pipe is $\Phi 216.3 \times 8.2$ (STKN400), and the following values are used for the parameters of the building oil damper.

- Damping force of brace-type oil damper F_T
- Displacement of brace-type oil damper u_T

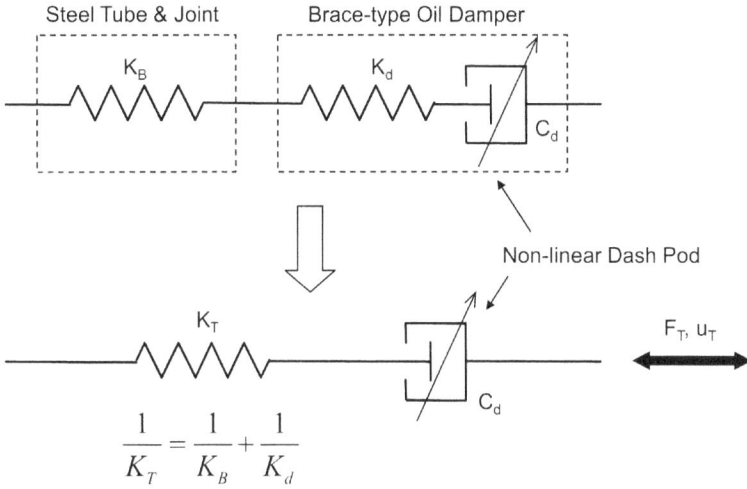

Figure 4.8 Brace-type oil damper analysis model.

- Compression rigidity of the oil damper part K_d = 1400 kN/cm
- Compressive rigidity of steel pipe joint KB = 2252 kN/cm
- Overall rigidity K_T = 864 kN/cm
- Damping coefficient C_d of oil damper part (1st order damping coefficient and 2nd order damping coefficient shall be the following values)
- First-order attenuation coefficient C_1 = 125 kN·sec./cm
- Secondary attenuation coefficient C_2 = 0.068 · C_1 = 8.5 kN sec./cm
- Release speed V_r = 3.2 cm/sec.

4.3.4 Comparison of experimental and analytical results

As a result of the experiment, the occurrence of buckling of the brace-type oil damper and the occurrence of resonance of the internal parts of the oil damper, which were of concern, were not confirmed, and the experiment was judged to have been successful.

In addition, to verify the validity of the analysis model, the response analysis at the time of sine wave input and seismic wave input was compared with the experimental values.

Figures 4.9 and 4.10 show a comparison of the experimental and analysis results of the hysteresis loop with damping force: F_T-displacement: u_T (period 1.5 sec., 4.0 sec.) when a sine wave is input. The input waves shown in Table 4.1 were used. The time step Δt in the analysis was 0.001 sec.

It can be said that the experimental results and the analysis results in the vibration experiment of the brace-type oil damper are in good agreement regardless of the vibration period and the amplitude.

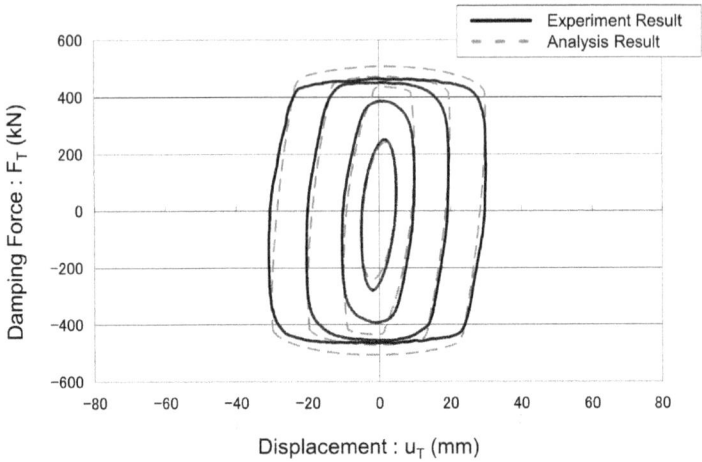

Figure 4.9 Damping force-displacement relationship (period: 1.5 sec.).

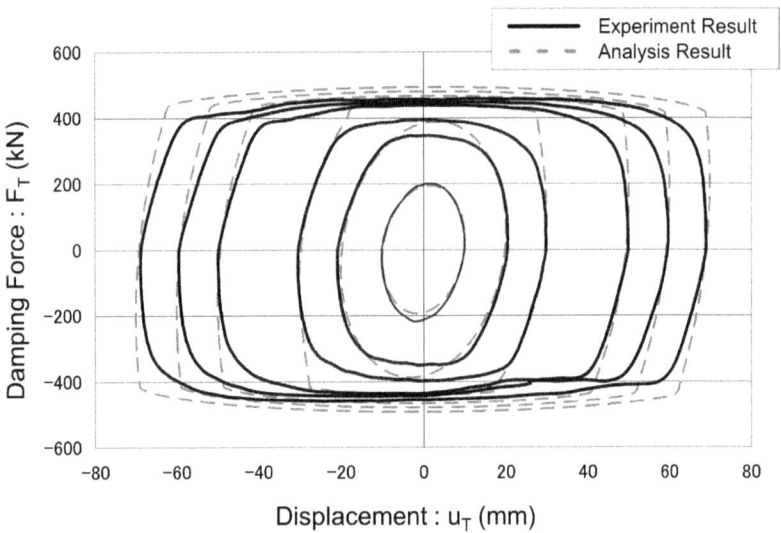

Figure 4.10 Damping force-displacement relationship (period: 1.5 sec.).

Comparison of experimental results and analysis results is shown in Figure 4.11 where the seismic wave of JMA 1995 KOBE NS (Kobe Marine Meteorological Observatory) was input to a steel-framed 21-story building with a maximum speed of 45 cm/sec. This figure indicates the sufficient accuracy of the result calculated by this analysis model.

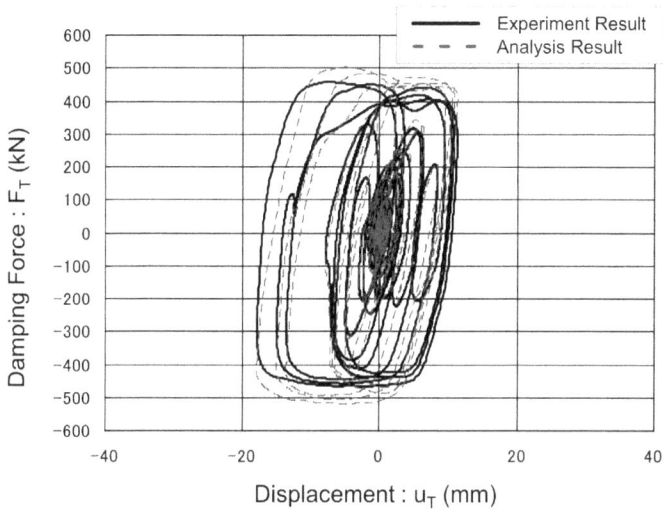

Figure 4.11 Damping force-displacement relationship (random wave).

4.3.5 Effect of steel pipe brace

To confirm the influence of the steel pipe brace part, it was decided to compare the experimental results without the steel pipe brace, which was conducted separately using the same oil damper as with the steel pipe brace. Figure 4.12 shows the damping force: F_T-displacement: u_T relationship with a period of 1.5 sec. and an amplitude (5.0 mm, 20.0 mm). There is no significant difference in the damping force-displacement relationship with this degree of difference in rigidity.

In addition, Figure 4.13 shows the relationship between the analysis results and K_T when the overall rigidity: K_T is reduced to about 1/2 to 1/4 for comparison. From this, it can be confirmed that when the rigidity of the Maxwell model becomes low, the loop of damping force-displacement relationship shows a thin ellipse.

From these results, it is considered important for the rigidity of the steel pipe brace portion to ensure high rigidity in consideration of the overall rigidity to fully exert the damping effect.

4.3.6 Required stiffness and bearing capacity of steel pipe brace

From Figures 4.12 and 4.13, it was confirmed that the rigidity of the steel pipe brace part is a major influential factor for the damping effect of the brace-type oil damper.

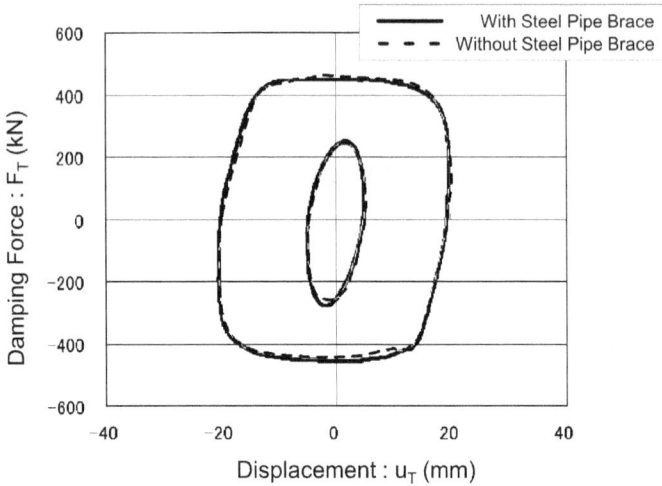

Figure 4.12 Damping force-displacement relationship (with and without brace).

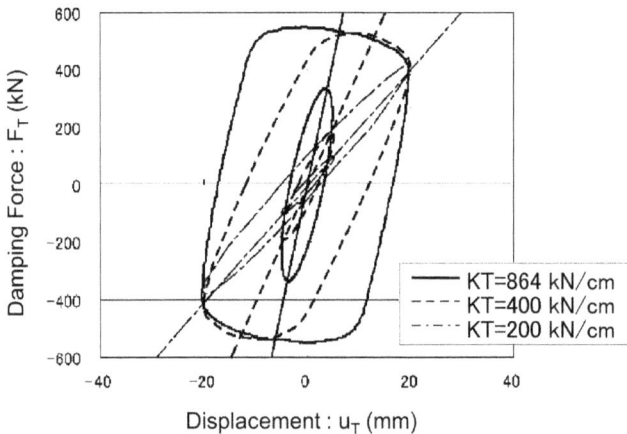

Figure 4.13 Damping force-displacement relationship (under different stiffness).

It is necessary to study the design in the building plan, but it was judged that a stable damping effect can be exhibited by ensuring the rigidity is higher than the rigidity confirmed by this experiment. Similarly, it was judged that it is desirable to secure the buckling resistance of the steel pipe brace part at least 1.5 times the maximum damping force of the oil damper.

4.4 OVERVIEW AND RESULTS OF FULL-SCALE FRAME DYNAMIC EXCITATION EXPERIMENTS

4.4.1 Purpose of the experiment

When the brace-type oil damper is attached to the frame as the actual expected length, the following points need to be examined.

(1) Behavior when mounted diagonally.
(2) Structural safety of each frame part and joint part by generated damping force.
(3) Analysis model considering frame.

Therefore, a full-scale frame dynamic vibration experiment was conducted for the purpose of confirming these.

At this time, the brace-type oil damper was used by attaching the test piece used in the full-scale dynamic vibration experiment as it was at an angle. The experiment was conducted in November 1999 in an environment with an average temperature of 13.6°C.

At the same time, we also confirmed the workability by attaching and detaching the brace-type oil damper to and from the frame.[6, 7]

4.4.2 Test specimen and experimental method

In the experiment to confirm the damping characteristics of the brace-type oil damper, the brace-type oil damper used in the full-scale dynamic vibration experiment in Figure 4.7 was attached to the full-scale frame as shown in Figure 4.14. Then, a full-scale frame dynamic vibration experiment was conducted.

The basic specifications of the full-scale test frame of this experiment are

- Floor height 4,000 mm.
- Span 7,000 mm.
- Beam (horizontal) material BH-800 × 350 × 16 × 36 (SM590 material).
- Pillar (vertical) material □ (Rectangular Shape) -900 × 300 × 25 × 36 (SM590 material).
- Mechanism added to prevent the frame from falling off the surface.
- Joint pinned with the floor.

As the force test device, the force test device owned by the Nuclear Power Generation Technology Organization was used.

The measurement items are as shown below.

- Force test device reaction force.
- Force test device displacement.
- Distortion generated at the joint and frame.

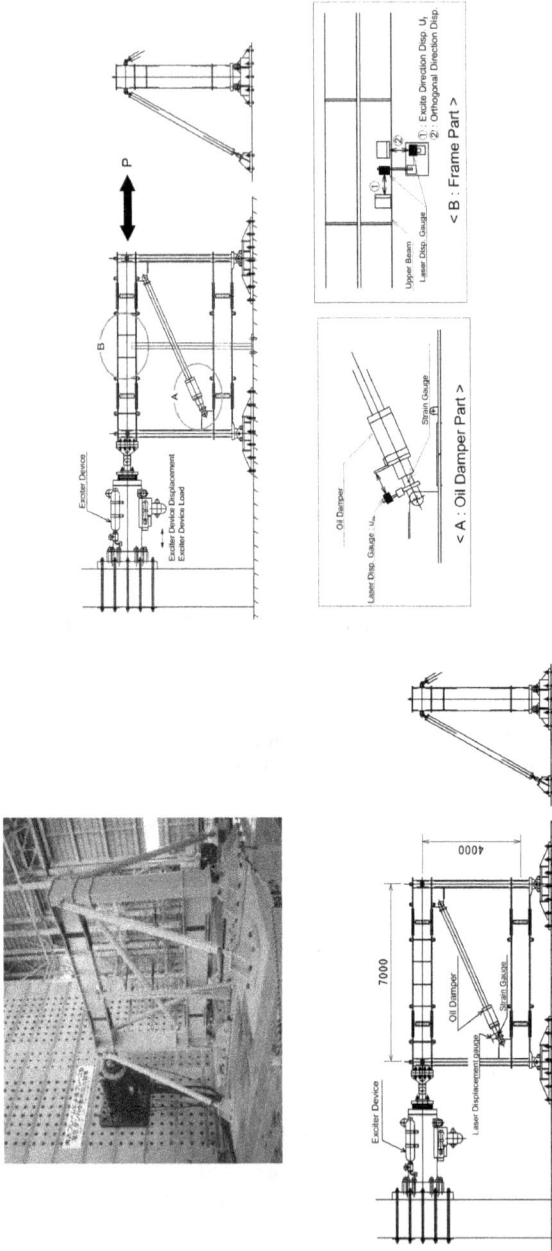

Figure 4.14 Full-scale frame excitation experiment plan.

Table 4.2 Test conditions

Amp. cycle	±5.0 mm	±10 mm	±20 mm	±30 mm	±40 mm
1.5 sec	○	○	○	○	○
4.0 sec	○	○	○	○	○

The frequency-dependent parameters in this experiment were set as shown in Table 4.2.

As for the frequency-dependent parameter setting, as in the full-scale dynamic vibration experiment, vibration is performed with the image of a steel-framed 10-story medium-rise office building and a steel-framed 40-story skyscraper office building. The period was set to 1.5 sec. and 4.0 sec., the span of the full-scale test frame was set to 7,000 mm, and the floor height was set to 4,000 mm.

The amplitude parameters were set within the range of 1/800 to 1/100 of the interlayer deformation angle of the full-scale test frame.

4.4.3 Analytical model for full-scale frame dynamic excitation experiment

For the analysis, a three-dimensional frame analysis program "RESP-T" (Structural Planning Institute Co., Ltd.) with the added function of the non-linear Maxwell model was used.

Here, regarding the brace-type oil damper, the rigidity K_B (2252 kN/cm) of the steel pipe brace part and the gusset plate part connected in series with the oil damper part is taken into consideration, and the oil damper part is a single unit. The overall rigidity K_T (864 kN/cm) including the compression rigidity K_d (1400 kN/cm) was calculated as a series spring, and the analysis model incorporated into the frame was used as shown in Figure 4.15. The joint is evaluated as a joint panel, and the basic characteristics of the oil damper are the primary damping coefficient C_1 (125 kN sec./cm) and the secondary damping coefficient C_2 (= 0.068 · C_1) is the relief speed V_r (3.2 cm/sec.). Table 4.3 shows a list of the specifications of the analysis model. In the numerical calculation, the analysis was performed by incorporating the brace-type oil damper as an additional damping force in the vibration equation as shown in the following equation as an element of the nonlinear Maxwell model. When incorporating, the term of the additional damping force (θ) due to this nonlinear Maxwell element was added as the element force in consideration of the mounting angle. For the calculation method, the method in Chapter 3 was used.

$$[P] = [M] \cdot \{\ddot{X}\} + [C] \cdot \{\dot{X}\} + [K] \cdot \{X\} + \left[F(x_{ik}, x_{ik})\right] \times \cos\theta \qquad (4.1)$$

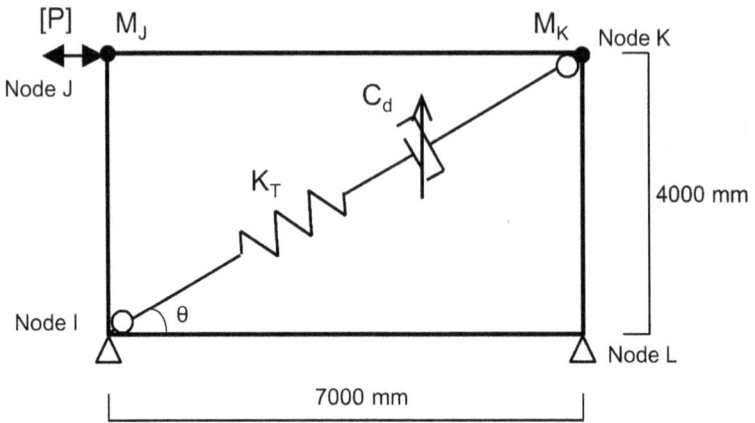

Figure 4.15 Analysis model.

Table 4.3 Analytical model parameters

Overall rigidity of braced oil damper	K_T	864(kN/cm)
1st damping coefficient of brace oil damper	C_1	125(kN/sec./cm)
2nd damping coefficient of brace oil damper	C_2	$0.068 \cdot C_1$
Braced oil damper relief speed	V_r	3.2(cm/sec.)
Young coefficient for columns and beam members	E	19600(kg/mm2)
Column member cross section		$\square$−900×300×36×25
Beam member cross section		H−800×350×16×36
Column and beam material		SM590
Cross-sectional area of the pillar AC	A_C	76200(mm2)
Column cross-section secondary moment	I_C	1.19×109(mm4)
Beam cross-section secondary moment	I_g	4.19×109(mm4)
Self-weight of column members (per pole)		23.9(kN)
Own weight of beam members (per beam)		20.4(kN)
Mass	Mj,Mk	2.26(kN·s2/ m)
Damping constants for columns and beam members		0.001(%)

Here,
   [P]: Excitation force
   [M]: Frame mass
   [C]: Frame internal viscosity damping coefficient
   [K]: Frame rigidity
   [F]: Additional damping force considering the mounting angle due to
        the nonlinear Maxwell element
   K_T: Overall rigidity of brace-type oil damper

C_d: Damping coefficient of brace-type oil damper
$\{X\}$: Response displacement of the frame

4.4.4 Comparison of experimental and analytical results

As a result of the experiment, it was confirmed that there was no problem in each part of the frame and the joint part due to the generated damping force that was a concern. Regarding workability, it was confirmed that the spherical bearing works effectively, and the attachment/detachment time is about 5 minutes at one place, and there is no problem in practical use as on-site work.

Figures 4.16 and 4.17 show a comparison of the experimental and analysis results of the frame excitation force: P-displacement: u_F (period 1.5 sec., 4.0 sec.) when a sine wave is input. Figures 4.18 and 4.19 show the comparison between the experimental results and the analysis results related to the damping force: F_d-displacement: um relationship only in the oil damper part in this experiment. The time step Δt in the analysis was 0.001 sec.

It can be said that the experimental results and analysis results of the vibration experiment with the brace-type oil damper attached to the full-scale frame are the same before and after the relief of the oil damper regardless of the vibration period and amplitude.

4.4.5 Comparison of energy consumption of the frame

Table 4.4 and Figure 4.20 show the calculation results of the energy consumption absorbed by the frame and brace-type oil dampers by experiments

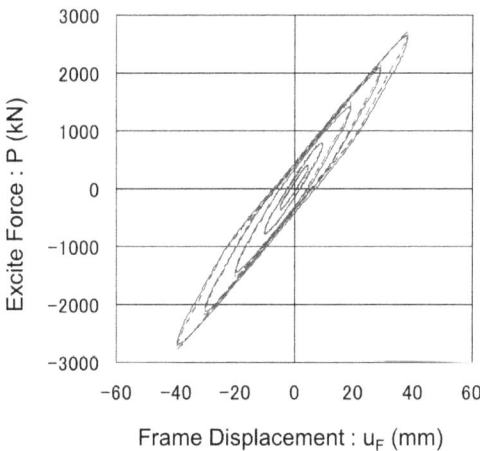

Figure 4.16 Excitation force-frame displacement relationship (period: 1.5 sec.).

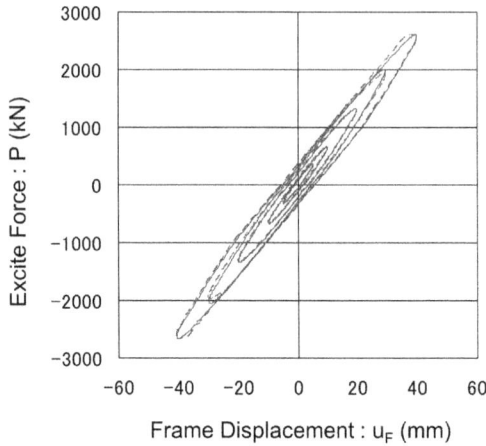

Figure 4.17 Excitation force-frame displacement relationship (period: 4.0 sec.).

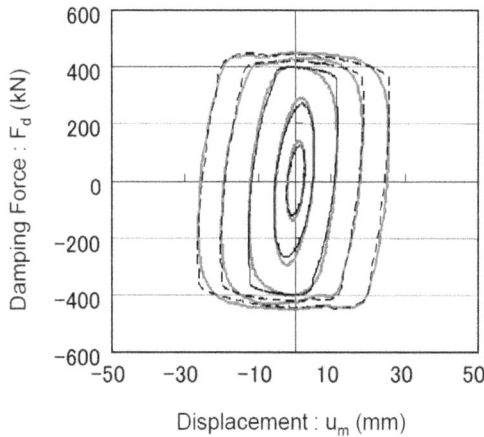

Figure 4.18 Excitation force-damper displacement relationship (period: 1.5 sec.).

from the loop area related to the excitation force and displacement of the frame. The frame with the brace-type oil damper consumes about 10 times more energy than the frame without the brace-type oil damper. In vibration period of 1.5 sec., maximum displacement of 30 mm and 40 mm, the energy consumption ratio (Ed/En) is small because the damping force of the damper enters the relief region. In vibration period of 4.0 sec., the small displacement vibration, since the vibration speed is small, the error can be confirmed to be slightly large, but the absolute value of the error is small. The energy consumption without a damper differs depending on the vibration cycle, and the energy consumption increases when the vibration cycle is fast. However, the effect on the vibration-displacement relationship

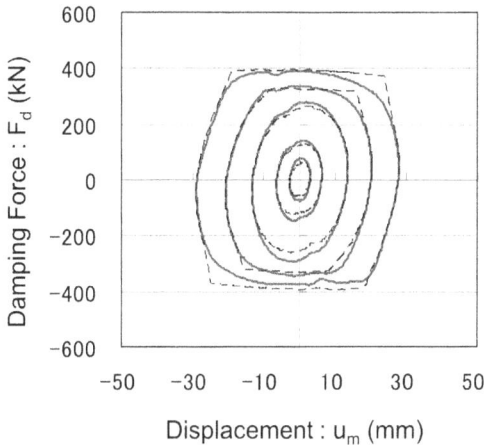

Figure 4.19 Excitation force-damper displacement relationship (period: 4.0 sec.).

Table 4.4. Comparison of frame energy consumption

	With damper		Without damper		
Cycle [sec.]	Max disp. [mm]	Ed [N.m]	Max disp. [mm]	En [N.m]	Ed/En
1.5	4.9	1.33×10^3	4.9	1.32×10^2	10.1
	9.9	5.30×10^3	10.0	4.65×10^2	11.4
	19.5	1.87×10^4	19.1	1.66×10^3	11.3
	29.7	3.64×10^4	29.6	4.71×10^3	7.7
	38.8	5.74×10^4	38.7	1.06×10^4	5.4
4.0	5.0	1.18×10^3	5.7	3.06×10^2	3.9
	9.9	3.50×10^3	9.9	5.78×10^2	6.1
	19.8	1.36×10^4	19.8	9.83×10^2	13.8
	29.5	2.80×10^4	29.8	2.20×10^3	12.7
	40.3	4.57×10^4	40.3	5.06×10^3	9.0

behavior with the damper shown in Figures 4.16 and 4.17 is small. From these facts, it was confirmed that the brace-type oil damper exerts a damping force from a small amplitude to a large amplitude.

4.4.6 Relationship between the displacement of the frame and the amount of strain at the joint and some points of the frame

Figure 4.21 shows the points with high stress levels without dampers for each part. This result is the result when the frame is pushed from the

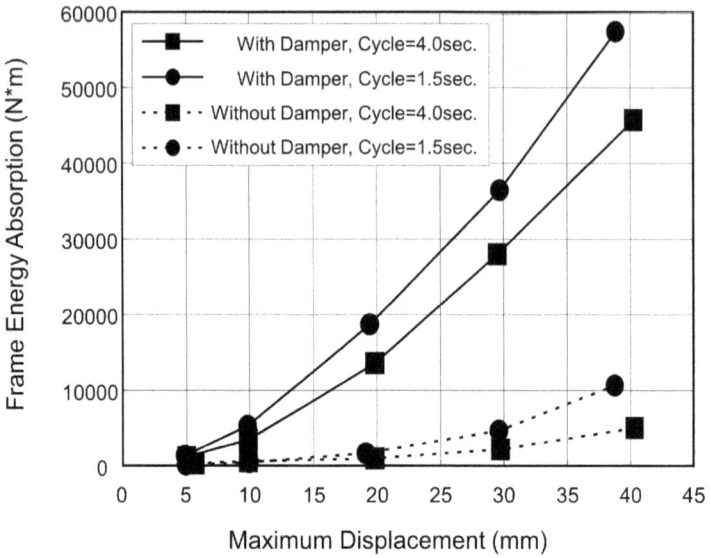

Figure 4.20 Frame energy consumption.

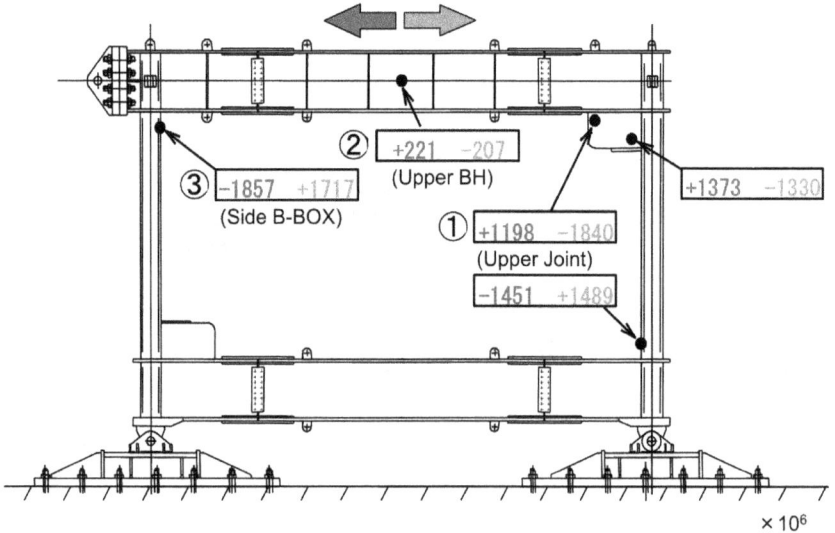

Figure 4.21 Max. distortion in each section (without damper).

left to the right in the figure. In addition, since the amount of generated strain depends on the amount of frame deformation regardless of the natural period, the experiment performed for 1.5 sec. 40 mm is shown as a representative.

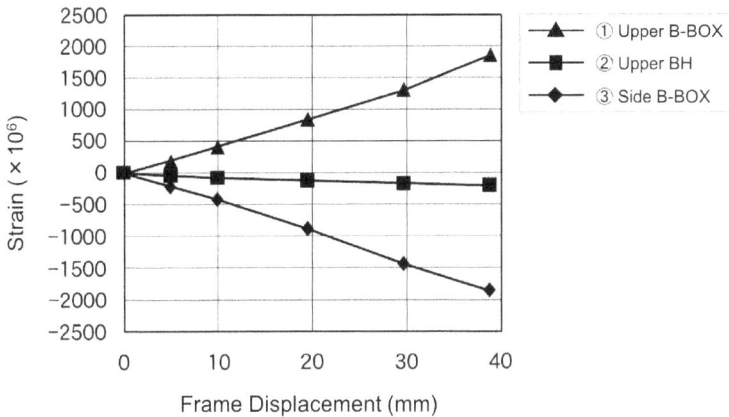

Figure 4.22 Distortion in each section (without damper).

Figure 4.22 shows the amount of strain generated in each part when vibration is applied until the frame displacement becomes about 40 mm in a period of 1.5 sec. with a damper.

As a result, since the upper beam portion only transmits the deforming force, the amount of generated strain is small. The joint and side columns are subject to tension, compression, and bending deformation at the same time, so the amount of strain is large, but each part maintains linearity for deformation up to 40 mm, and the amount of strain is elastic. It was confirmed that it was within the range.

4.5 SUMMARY

We conducted a full-scale dynamic vibration experiment and a full-scale dynamic frame vibration experiment of a brace-type oil damper that can be directly attached to the building frame in a shape like a brace for the purpose of high damping of the building. A damping experiment was conducted to confirm the damping characteristics. The main findings obtained are as follows.

(1) In a full-scale dynamic vibration experiment, buckling and resonance of internal parts of the oil damper were not confirmed in the brace-type oil damper, if it has this length and rigidity, even if a steel pipe brace part is attached. It was confirmed that there was no problem.

(2) For the analysis modeling of the brace-type oil damper part, the element that modeled the oil damper part with the Maxwell model and the element that modeled the axial rigidity of the brace part and the end joint part with the shaft spring. A model combined in series

confirmed accurate agreement with the experimental results in both sinusoidal and random wave excitation.

(3) In the full-scale frame dynamic vibration experiment, it was confirmed that the influence of the generated damping force on each part of the frame and the joint part does not cause any problem.

(4) It was confirmed that the spherical bearing works effectively in construction and the attachment/detachment time is about 5 minutes at one place, which does not cause any problem in practical use.

(5) Regarding the analysis modeling incorporated in the frame, confirmed accurate agreement experimental result with analysis result using the model modeled by the Maxwell model, which considers the influence of the axial rigidity of the peripheral members.

REFERENCES

1. Toshiko Okuzono, Osamu Takahashi, Masayuki Ninomiya: "Structural planning institute building; vibration control structure by directly attaching oil damper", *Steel Technology*, pp.12-1–12-6, 1999.3.

2. Osamu Takahashi, Yohei Sekiguchi: "Oil damper analysis algorithm and subrutin using Maxwell model", *Passive Vibration Control Structure Symposium 2001*, 2001.12.

3. Osamu Takahashi, Youhei Sekiguchi: "Time-history analysis model for nonlinear", SEWC2002, 2002.10.

4. Naofumi Igarata, Fumiya Iiyama, Kazuhiko Shibata, Toshifumi Okuzono, Osamu Takahashi, Masaki Shibata, Yasushi Miyamoto: "Development of hydraulic damper with load column for minute amplitude-Part 1 Single dynamic force test", *Japan Architectural Institute of Japan Academic Lecture Abstracts (Tohoku)*, B-2 Volume, pp.847–848, 2000.9.

5. Masaki Shibata, Toshifumi Okuzono, Osamu Takahashi, Yasushi Miyamoto, Fumiya Iiyama, Naofumi Igarata, Kazuhiko Shibata: "Development of hydraulic damper with load column for minute amplitude-Part 2 Single dynamic loading test (2)", *Architectural Institute of Japan Conference Academic Lecture Abstracts (Tohoku)*, B-2 Volume, pp.849–850, 2000.9.

6. Fumiya Iiyama, Naofumi Isobata, Kazuhiko Shibata, Toshifumi Okuzono, Osamu Takahashi, Yasushi Miyamoto, Masaki Shibata: "Development of hydraulic damper with load column for minute amplitude-Part 3 Dynamic loading test using full-scale frame", *Architectural Institute of Japan Conference Academic Lecture Abstracts (Tohoku)*, B-2 Volume, pp.851–852, 2000.9.

7. Yasushi Miyamoto, Masaki Shibata, Osamu Takahashi, Toshifumi Okuzono, Fumiya Iiyama, Naofumi Iwahata, Kazuhiko Shibata: "Development of hydraulic damper with load column for minute amplitude-Part 4 Dynamic loading test using full-scale frame②", *Architectural Institute of Japan Conference Academic Lecture Abstracts (Tohoku)*, B-2 Volume, pp.853–854, 2000.9.

Chapter 5

Evaluation of vibration damping performance of actual building using brace-type oil damper and confirmation experiment

5.1 INTRODUCTION

In recent years, for the purpose of improving safety against earthquakes and improving comfort against environmental vibrations, a vibration-damping structure system that attaches a viscous damper to a building as a vibration damping device to make the building highly damped has been attracting attention.[1, 2]

Vibration-damping devices that use oil dampers have the following merits.

- The temperature dependence is negligible.
- The devices have been in use for many years as a vibration control member for transportation machinery such as railways and automobiles.
- The devices are easy to model and solve in analysis.

In addition, the authors have described a brace-type oil damper as an application example of a vibration-damping device using a building oil damper.

(1) Make sure that the damping force does not adversely affect the structural frame.
(2) Make it easy to remove during construction.

To satisfy these two points, we have developed a structure in which an oil damper is installed in series in the brace and can be directly attached to the building structure via a pin joint that can be joined in the field.

Here, we have had the opportunity to install the brace-type oil damper developed based on this development concept as the first practical use in an actual building, so we will report an example of its design and utilization.

In addition, since we were able to obtain an opportunity to confirm the damping effect by the method shown below at the time of construction and

DOI: 10.1201/9781003290261-6

after completion of this building, we will also summarize and report on the results.

- Vibration machine experiment (immediately before delivery of building completion).
- Human-powered vibration experiment (two years after the building is completed).
- Seismic observation (at the time of the Chuetsu Earthquake on October 23, 2004).

5.2 BUILDING DESIGN POLICY AND OVERVIEW OF BRACE-TYPE OIL DAMPER

5.2.1 Building design concept

In this property, we decided to adopt the newly developed brace-type oil damper of the viscous vibration-damping device as the first practical use.

From the standpoint of architectural design for the purpose of collaboration between design and structural design, this property is equipped with a vibration-damping device that can simultaneously achieve the following three issues to improve the building performance of the target building.

- Improved livability against daily wind and traffic vibration.
- Improved safety against earthquakes.
- Development of a system integrated with the design (dampers to show).

Focusing on these issues, the required performances considering the dampers are

(1) Maintenance-free equipment that basically does not require replacement.
(2) A device that is not affected by temperature because it may be installed on the window side.
(3) A device with reproducible performance and easy analysis modeling.
(4) A device that can exert effects from small amplitude to large amplitude.
(5) A device that can be arranged so that the damping force does not adversely affect the structural frame.

In addition, as the design criteria of the building, we aimed at the level where the response of the building is reduced to about 1/2 by installing this damper.

5.2.2 Outline of the structure of the brace-type oil damper

Figure 5.1 shows the structural outline and layout of the brace-type oil damper installed in this building.

Basically, it is a structure in which the oil damper part and the steel pipe brace part are connected in series. The oil damper part is a 500 kN type building oil damper (outer diameter = φ200) and the steel pipe brace part is a building steel pipe connected in series. At the end, a spherical bearing is adopted, and a mounting pin is inserted into it so that each part can function normally even when the workability of on-site mounting and the force in the out-of-plane direction are applied as shown in Figure 5.2. By using a mechanical pin with a spherical bearing that combines this taper sleeve and taper pin, the mechanism is such that the mechanical gap is 0.1 mm or less during on-site construction.

In consideration of the processing accuracy of the pin, the gusset plate and steel frame machined by a steel frame processing company were requested to process the gusset plate attached to the skeleton frame as a set of vibration-damping members. After welding and joining the skeleton and constructing it at the construction site, the part where the oil damper part and the steel pipe braces that were separately brought into the construction site are connected in series and the gusset plate part are connected at the construction site. In addition, regarding the amplitude stroke, it was set to ± 80 mm in consideration of the construction error and the followability up to the interlayer deformation angle of 1/50 on the premise that it is installed as a vertical brace.

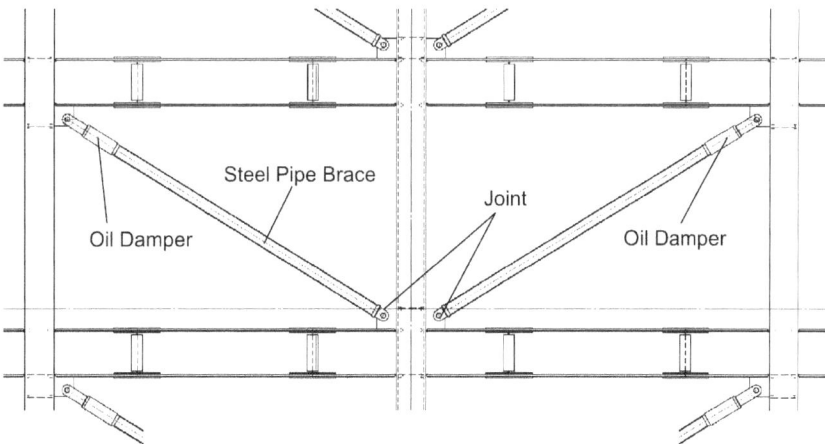

Figure 5.1 Brace-type oil damper installation.

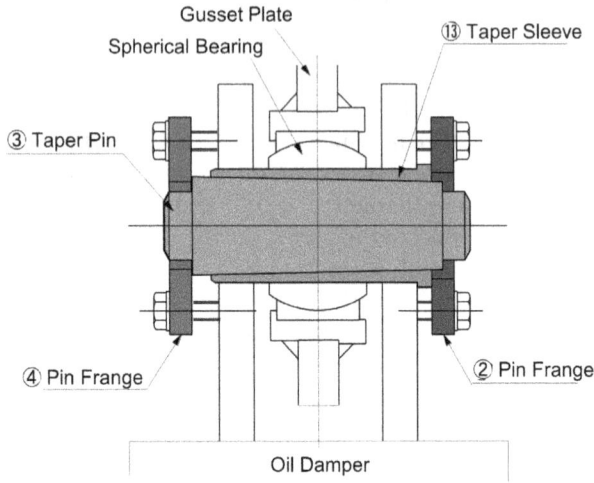

Figure 5.2 Joint.

5.2.3 Basic characteristics of brace-type oil damper

Figure 5.3 shows the basic characteristics of the brace-type oil damper installed in this building. The basic characteristics can be expressed by the relationship between the damping force and the speed of the dashpot in the damper. The building oil damper has a bilinear characteristic in which

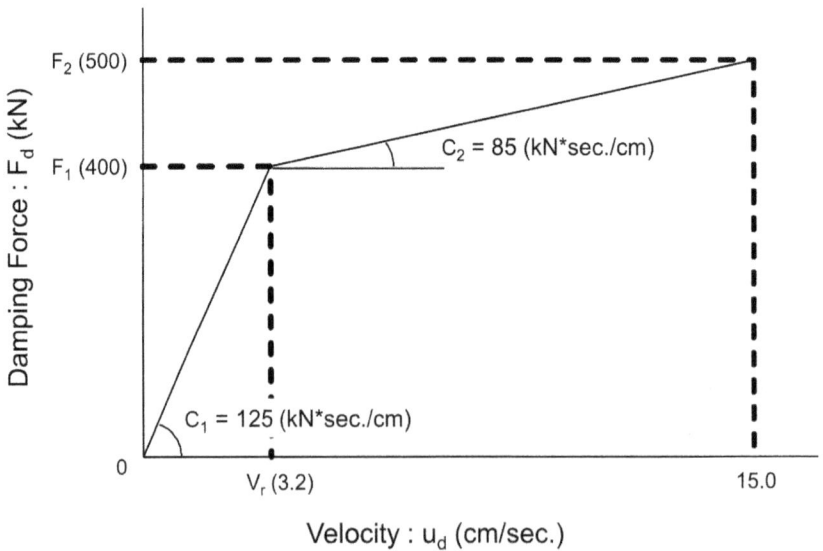

Figure 5.3 Damping characteristics of dashpot.

the release valve is opened, and the damping force is suppressed for a load exceeding a predetermined value by providing a release valve in consideration of fail safety. The damping characteristic at that time is set to the maximum damping force: F_2 (500 kN), and after exceeding the primary damping coefficient: C_1 (12.5 t · sec/cm) and the release damping force: F_1 (400 kN), the second-order attenuation coefficient C_2 is defined to be 0.068 times of F_1. These are to the extent that the damping force of the building oil damper is generally set to be less than the relief damping force at the seismic motion level that occurs rarely with reference to the design example of the vibration-damping building, and more than the relief damping force at the seismic motion level that occurs extremely rarely.

5.2.4 Performance verification experiment

In the experiment to confirm the damping characteristics of the building oil damper, a dynamic vibration test was conducted using the vibration test equipment as shown in Figure 5.4. The damping force was measured with a load cell and the displacement was measured with a laser displacement meter.

In addition, the frequency-dependent parameters in this experiment were set and confirmed as shown in Table 5.1, including the primary natural period of the target building (1.25 sec.).

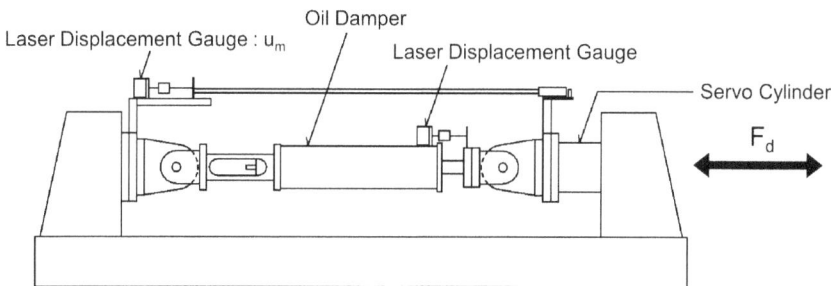

Figure 5.4 Exciter device.

Table 5.1 Excitation period and amplitude list

Cycle [sec.]	1.5	1.25	2.0
Shake amp. [mm]	10.0	10.0	10.0
	20.0	20.0	20.0
	30.0	30.0	30.0
	—	—	40.0

5.2.5 Comparison of experimental and analytical results

Figure 5.5 shows the experimental results of the damping force-displacement hysteresis loop, which shows the basic characteristics of the building oil damper. As a result, the shape of the hysteresis loop obtained from the excitation experiment shows an elliptical shape peculiar to the velocity-dependent damper, and it is considered that the simple dashpot model is insufficiently modeled. In addition, while the damping force at the maximum displacement is also amplitude-dependent, the displacement at the maximum damping force does not show a large fluctuation, so it is considered that the behavior can be modeled by the Maxwell model.

Here, the compression rigidity of the oil damper portion is set to K_d (= 1400 kN/cm) as the setting performance of the 500 kN type oil damper.

Figure 5.5 shows the damping force-displacement curve comparing the experimental results and the analysis results, and Figure 5.6 shows the damping force: F_d-velocity: u_d relationship. The velocity of the experimental results was plotted in um.

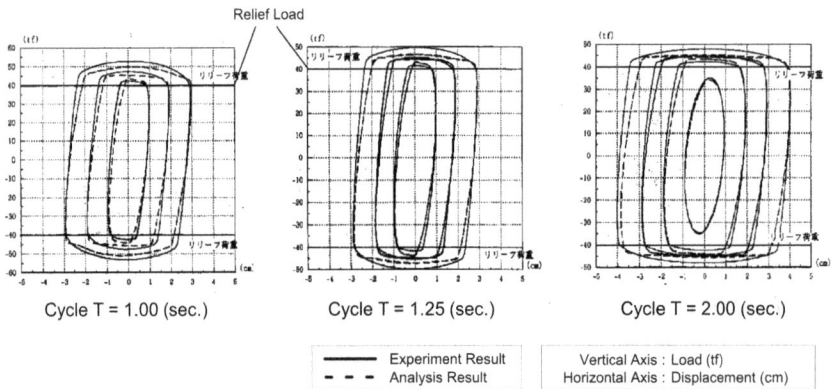

Figure 5.5 Damping force-displacement relationship.

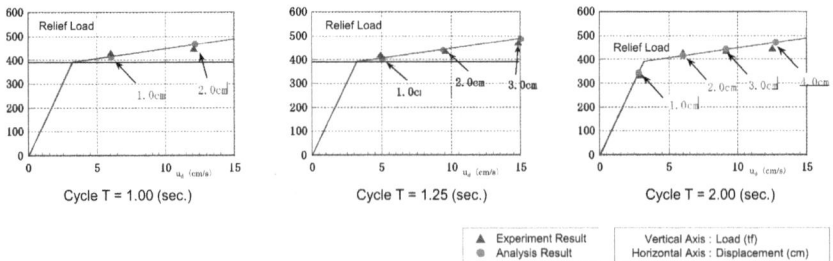

Figure 5.6 Damping force-velocity relationship (experiments and analysis).

5.3 OUTLINE OF BUILDING AND STRUCTURAL PLAN

5.3.1 Building outline

The building[3, 4] shown in Figure 5.7 is an office building with nine floors above ground and two floors below ground, which was constructed in Nakano-ku, Tokyo. A total of 72 dampers are arranged, one set at a time, and the exterior is constructed by continuously incorporating a light "V"-shaped form in the height direction by actively incorporating it as a design element in both the external space and the internal space. Table 5.2 shows the overview for the building.

5.3.2 Structural plan outline

Figure 5.8 shows the reference floor plan, the representative framework diagram, and the layout of the brace-type oil damper.

The above-ground part is made of steel to make the office space wider, and the underground part is made of reinforced concrete and steel-framed reinforced concrete. The outer wall is a PC (Precast Concrete) plate panel and a glass curtain wall, and these have a structure having deformation followability that hardly affects the rigidity of the building.

The brace-type oil damper, which is a vibration-damping device, was placed at the position shown in the reference floor plan of Figure 5.10 on each of the nine floors above the ground. A total of 8 units, 4 in the X direction and 4 in the Y direction, are arranged, and 72 units are installed in the entire building.

Figure 5.7 Building exterior and brace-type oil damper installation.

Table 5.2 Building overview

Location	Nakano Ward, Tokyo
Total floor area	7,238.82 (m²)
Number of stories	9 floors above ground, 2 floors below ground
Structural format	Steel structure (above ground)
Damping device	Brace-type oil damper (max. damping force 500kN type 72 units)

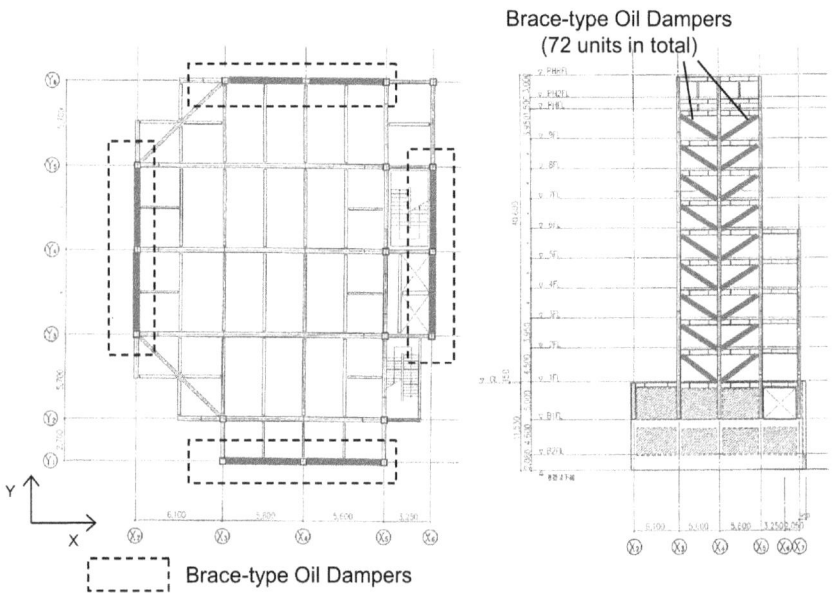

Figure 5.8 Building exterior and brace-type oil damper installation in the standard floor plan and the representative section.

In addition, as a construction procedure, a brace-type oil damper was built after sufficient curing for steel frame construction at the same time as the steel frame construction of steel columns and steel beam members.

Figure 5.9 shows the layout at the time of construction, showing the situation where the brace-type oil damper was built before the floor slab concrete was placed. In addition, it can be confirmed that the construction time of the end construction pin can be joined by the builder of the steel frame processing company in about 5 minutes per place, and the removal can be done in about 5 minutes per place as well.

Figure 5.9 Building exterior and brace-type oil damper installation.

5.4 BUILDING VIBRATION ANALYSIS OVERVIEW

5.4.1 Analysis model

For the vibration analysis of the building, the above-ground part was modeled as a nine-mass equivalent bending shear type with one layer and one mass, and the viscosity inside the building was proportional to the initial rigidity (h = 0.02). Table 5.3 shows the natural period of the building. The model of the brace-type oil damper was evaluated by the Maxwell model based on the damping characteristics confirmed in the dynamic vibration experiment. Here, as shown in Figure 5.10, the model of the brace-type oil damper is that the overall rigidity is K_T (= 900 kN/cm) and the rigidity of the steel pipe part and the joint part is K_B (= 2600 kN/cm), and the compression rigidity of the oil damper part: K_d (= 1400kN/cm) is combined in series. Figure 5.11 shows the analysis of brace-type oil damper.

5.4.2 Analysis results

Figure 5.12 shows the results of building response analysis in the X direction when the TAFT EW 1956 wave was input. Here, the maximum velocity of the input seismic motion is standardized to the following values.

- LV1: 25.0 cm/sec.
- LV2: 50.0 cm/sec.

In addition, the response values when the brace-type oil damper is not installed (h = 0.02) and when the viscosity damping inside the building

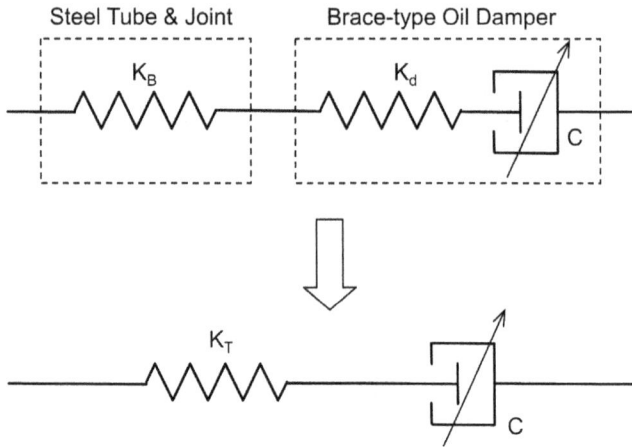

Figure 5.10 Brace-type oil damper analysis model.

Table 5.3 Building-specific period list

Y-direction 1st mode	1.38 (sec.)
X-direction 1st mode	1.24 (sec.)
Twist direction 1st mode	1.01 (sec.)
Y direction 2nd mode	0.47 (sec.)
X direction 2nd mode	0.44 (sec.)
Twist direction 2nd mode	0.36 (sec.)

is changed to h = 0.04, 0.06, 0.08, 0.10 are shown at the same time for comparison.

From the response results, a high damping effect of h = 0.10 for equivalent damping in LV1 and h = 0.08 for equivalent damping was observed in LV2 for both the maximum response layer shear force of the skeleton and the maximum response interlayer displacement. The reason why there is a difference in the damping effect between LV1 and LV2 is that the brace-type oil damper responds effectively in the region of high damping coefficient before release at the time of LV1 response, and the response speed becomes faster in LV2. This is because it responds effectively in the region of low damping coefficient after release.

The final value of energy was as follows.

- Input energy 3.5 × 104 (kN m).
- Attenuation energy 3.9 × 103 (kN m).
- Oil damper 3.0 × 104 (kN m).

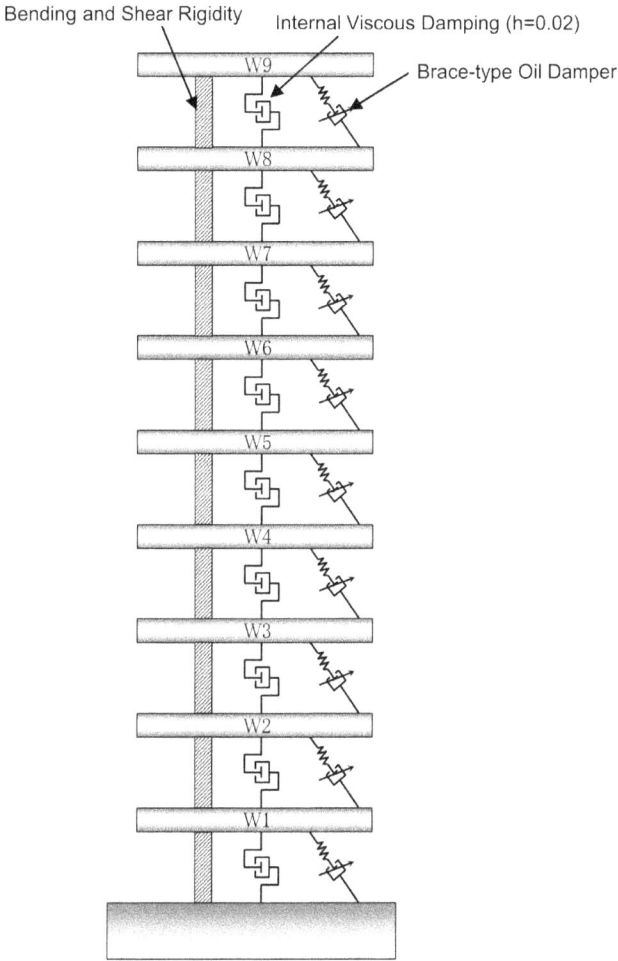

Figure 5.11 Brace-type oil damper analysis.

Here, it is assumed that all the energy absorbed by the oil damper is converted into the temperature rise of the oil damper, and the temperature rise
is calculated.

The specific heat of the oil damper for the temperature rise was tenta-
tively adopted as the specific heat of iron.

- Energy absorption per oil damper

    3.0×104 (kN · cm)/4 (books) = 7.6×103 (kN · cm) · · · · · · · ①

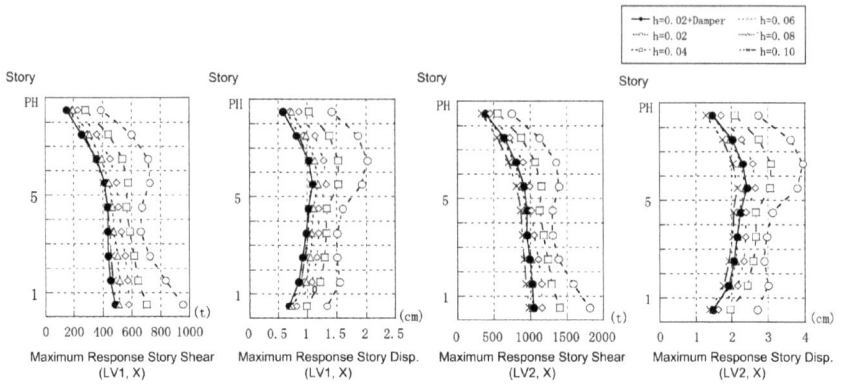

Figure 5.12 Results of building response analysis.

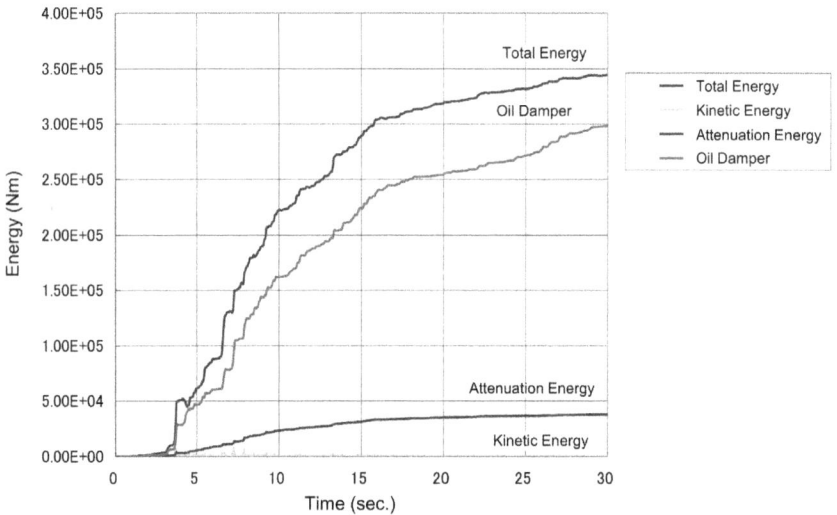

Figure 5.13 Brace-type oil damper analysis.

- Mass per oil damper

  Oil damper (main body) 1.5×101 (kg) $\cdots\cdots$ ②

- Temperature rise per oil damper

  Q (cal) = m (g) × c (cal/g · k) × T (k) $\cdots\cdots\cdots\cdots\cdots$ ③

  Calorie Q: 1.00 (kN · cm) = 2.39 (cal)

7.6 × 103 (kN · cm) = 1.82 × 104 (cal) · · · · · From ①

Mass m: 1.5 × 104 (g) · · · · · From ②

Specific heat (iron) c: 0.105 (cal/g · K)

Temperature change T: T (K)

From formula ③

T = Q ÷ (m × c)

∴ 1.82 × 104 (cal) ÷ (1.5 × 104 (g) × 0.105 (cal/g · k)) = 11.5 (k)

Based on the above calculation, it is expected that the temperature will rise by 11.5°C per oil damper. Figure 5.13 shows the amount of energy absorption of brace-type oil damper at the 1st floor.

5.5 PERFORMANCE CONFIRMATION BY VIBRATION MACHINE EXPERIMENT

Immediately before delivery of the building (December 1998), a steady vibration experiment using a vibrator was conducted in this building, and the control was performed by comparing the presence and absence of a brace-type oil damper based on the natural vibration characteristics and resonance curve. We have conducted an experiment to confirm the vibration effect and report the results.[5]

5.5.1 Outline of experimental plan

The vibration experiment was carried out in about one week with the building almost completed using an unbalanced type of vibration machine. The position of the oscillator was set on the rooftop floor (RFL) shown in Figure 5.14. Regarding the installation, it was attached to a frame made of H-shaped steel and fixed to the center of gravity of the building with an anchor. The outline of the oscillator is shown in Figure 5.15, and the performance specifications are shown in Table 5.4. The vibration direction was one horizontal direction on the east-west axis, and the vibration device output and the acceleration of each floor were measured. Table 5.5 shows the vibration cases.

The experimental case has a vibration-damping device (with a brake type oil damper attached) and no vibration-damping device (with the pin at the lower joint of the brake type oil damper removed and placed on the floor). In both cases, the experiment was carried out in the target period zone centered on the primary natural period and the secondary natural period of the building.

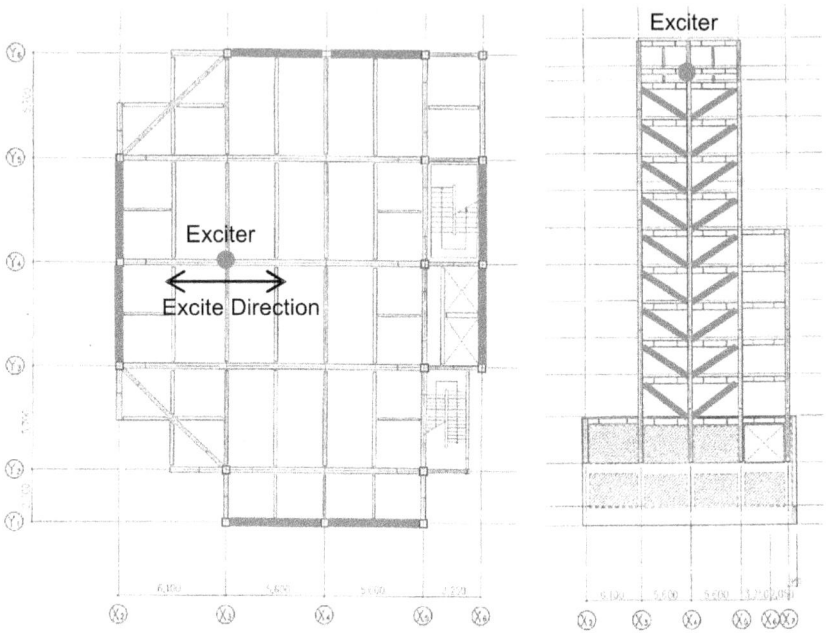

Figure 5.14 Exciter installation.

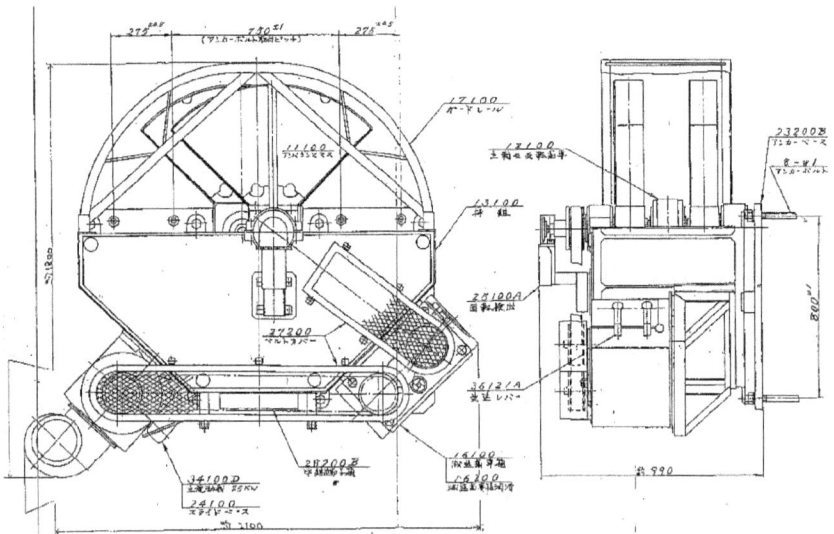

Figure 5.15 Exciter summary.

Table 5.4 Exciter performance specifications

Model	Horizontal plane rotational parallel biaxial unbalanced weight inverting formula (BCS-A-200DL Type)
Excite direction	Horizontal right angle bidirectional
Excite moment	2–10 kg.m (2 kg.m steps)
	10–200 kg.m (10 kg.m steps)
Max. excite force	5,000 kgf
	The first and second speed is 2.49 Hz or more
	3rd speed is 11.14 Hz or more
Excite waveform	sine wave
Frequency	0.2–20 Hz
Total weight	Approximately 2,100 kg
	(Body 700 kg, weight 700 kg, anchor base 300 kg, etc.)
Size	Approximately 2,100 × width depth 1,800 × height 1,000 mm
Power	AC3φ,3W,200,220V,50/60 Hz
	Input current 43–71A
	Start-up current 105A
Maker	Ito Seiki Co., Ltd.

Table 5.5 Excitation case

Case	Excite moment	Excite frequency	Damper installation
1	2 kN·m	0.7·2.4Hz	Without
2	1 kN·m	2.4·3.5Hz	Without
3	2 kN·m	0.7·2.4Hz	With
4	1 kN·m	2.4·3.5Hz	With

5.5.2 Experimental results

Table 5.6 shows the natural period obtained by the shaker experiment and the damping constant calculated by the $1/\sqrt{2}$ method. When determining the natural period and attenuation constant, the time when the maximum response amplitude occurs on the top floor (RFL) is used as the reference. Figure 5.16 shows the vibration mode.

Figure 5.17 shows the resonance curve plotting the response amplitudes of the top floor (RFL) and middle floor (4FL). A comparison was made between experiment results with and without vibration-damping device. Even at a very small amplitude level, the response displacement amount with vibration-damping device is about one-third compared to without the device in each period-band of both first and second mode.

Table 5.6 Natural period and damping constant

Damper	Vibration mode	Unique period [sec.]	Max. response amp. at RFL [cm]	Damping constant
Without damper	1st	0.971	0.348	0.014
	2nd	0.335	0.054	0.023
With damper	1st	0.943	0.118	0.039
	2nd	0.313	0.023	0.061

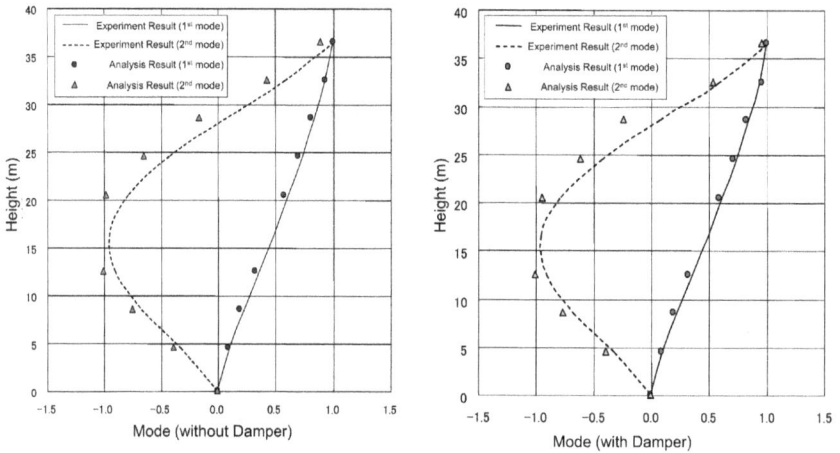

Figure 5.16 Eigen mode diagram.

5.6 PERFORMANCE VERIFICATION BY HUMAN-POWERED VIBRATION EXPERIMENT

In this building, two years after its completion (June 2001), we constantly measure fine movements and perform human-powered vibration by moving the weight of people on the rooftop floor of the building to measure natural vibration characteristics and damping constants due to free vibration. The result of this experiment is reported as follows.[6]

5.6.1 Outline of experimental plan

On the rooftop floor of the building, vibration was performed by moving the weight with a person, and the free vibration after the vibration was stopped and the load and displacement generated in the brace-type oil damper attached to the building were measured. Figure 5.18 shows the scene of human-powered excitation experiment.

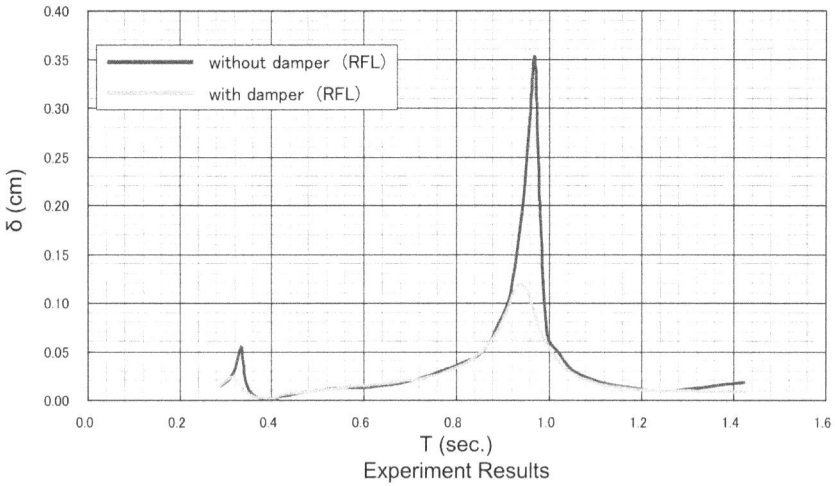

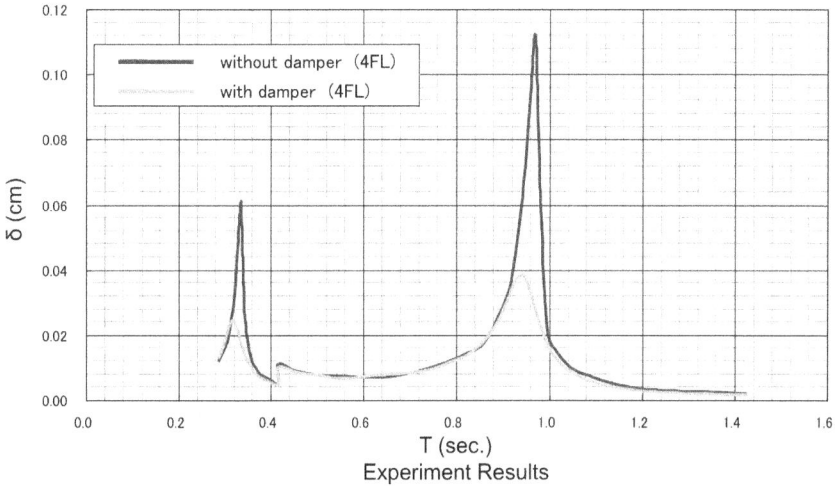

Figure 5.17 Resonance curve.

The procedure for manual vibration is shown below.

(1) Approximately 20 people will be lined up on the roof of the building in the vibration direction (east-west direction).
(2) The building is vibrated horizontally by each person moving like weight shift from one foot to other foot alternately according to the signal from the conductor and metronome which BPM is adjusted with the primary natural period of the building.
(3) Continue the vibration while monitoring the waveform, and when the amplitude stabilizes, stop the weight transfer, stop the vibration, and vibrate freely.

Figure 5.18 Human-powered excitation experiment.

The vibration is in one horizontal direction (east-west direction), and the vibration cycles are 0.87 sec. (constant tremor predominant cycle measured on the eve of the human-powered vibration experiment in this building), 0.94 sec. (natural period of the target building), and 1.00 sec. and carried out three times respectively.

The data was recorded on the data recorder from the start of vibration until the vibration of the building subsided after the vibration stopped. In addition, the vibration period was managed using a spectrum analyzer.

In addition, for the constant fine movement measurement conducted on the eve of the human-powered vibration experiment, acceleration vibration was measured, and the data was sent to the data recorder from 2:00 am to 4:00 am (about 120 min.).

5.6.2 Experimental results

The natural period and damping constant were calculated from the free vibration waveform of the 9th-floor ceiling accelerometer obtained in the experiment. The weather at the time of the experiment was fine, the wind speed was 1.0 m/sec. or less which is assumed to be ignorable.

The natural period was calculated as the reciprocal of the natural period calculated by reading the interval at which the wave passes the zero point.

The attenuation constant was calculated from the logarithmic decrement by reading the amplitude of the five waves after the vibration was stopped. The formula is shown below.

$$\delta = \frac{1}{N} \cdot \log_e\left(\frac{A_0}{A_N}\right) : \text{Logarithmic decrement}$$

A_0: Initial amplitude
A_N: Nth wave amplitude
N: Number of waves

$$h = \frac{\delta}{2\pi} : \text{Attenuation constant}$$

Table 5.7 shows the damping constant and the frequency during free vibration. Since the natural frequency at the time of attenuation fluctuates as the amplitude decreases, the average value is used for every three waves after the vibration is stopped.

The free decay waveform is shown in Figure 5.19.

The damping constant calculated from the logarithmic decrement by reading the amplitude of free vibration even at a very small amplitude level such as human-powered vibration is h = 0.04 to 0.05, and the micro-vibration level of a general steel structure is about h = 0.01. With this, it was confirmed that the main building equipped with the brace-type oil damper was highly attenuated.

Table 5.7 Damping constant

Shake frequency [Hz] Setting value		Ave. acceleration during excitation at ceiling on 9F [cm/s²]	Free Frequency [Hz]	Damping Constant [%]
1.15	1st	0.984	1.10	4.93
	2nd	1.016	1.10	4.17
	3rd	1.037	1.10	4.41
	Ave	1.012	1.10	4.50
1.06	1st	1.682	1.05	4.53
	2nd	1.796	1.05	4.31
	3rd	1.911	1.04	4.40
	Ave	1.796	1.05	4.41
1.00	1st	2.467	1.03	4.19
	2nd	2.552	1.03	4.43
	3rd	2.519	1.03	4.31
	Ave	2.513	1.03	4.31

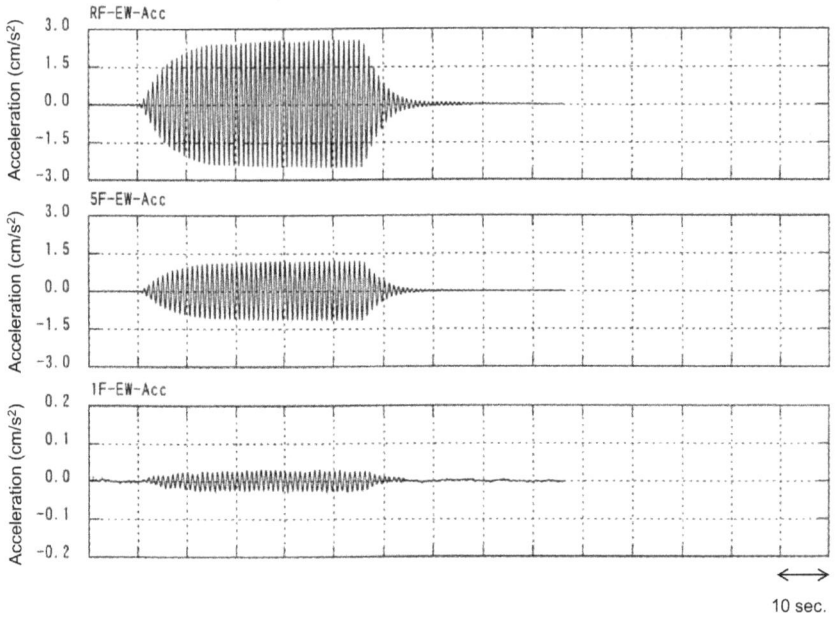

Figure 5.19 Input free vibration waveform.

5.7 PERFORMANCE CONFIRMATION BY SEISMIC OBSERVATION

In this building, long-term observations are being conducted to confirm the damping effect of the brace-type oil damper against actual disturbances such as earthquakes and winds. Here, we report the seismic observation results observed at this building when an earthquake with a magnitude of 6.8 and a seismic intensity of 6 or higher occurred in the Chuetsu area of Niigata Prefecture on October 23, 2004.

5.7.1 Outline of the observation plan

Figure 5.20 shows the outline of the observation plan of this building.[2, 7] Inside the building, seismographs are installed at RFL, 9FL, 5FL, 1FL, and pressure plate positions, and displacement meters and load meters are installed on the brace-type oil dampers on the 9th, 5th, and 1st floors. In addition, a seismograph is buried outside the building at the GL level, the flooring level of the building (GL-15.0 m), and the level with an N value of 50 or more (GL-50.0 m, Vs = 342 m/sec.), that is triggered by 2.0 cm/sec.[2] or more acceleration at GL-50.0 m and the data is recorded in the measuring instrument on the first basement floor. In addition, a wind direction and anemometer are also installed on the rooftop floor, and the system is

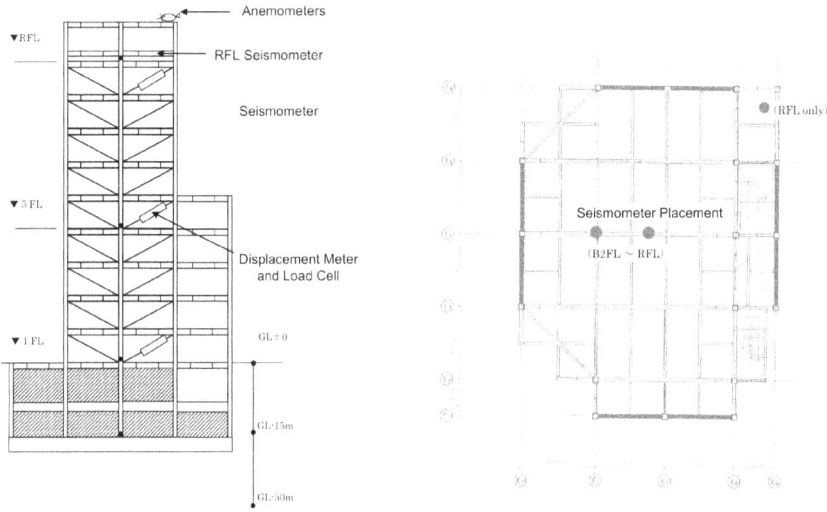

Figure 5.20 Measurement instrument installation.

such that all data is recorded in the same way as the seismometer, with the trigger of 15.0 m/sec. instantaneous wind speed. For data recording, the sampling interval during seismic observation is set to $\Delta t = 0.01$ sec., and the sampling interval during wind observation is set to $\Delta t = 0.1$ sec.

5.7.2 Seismic observation results

The results of the seismic response observed at this building when the Niigata Chuetsu Earthquake (magnitude 6.8, seismic intensity 6 upper) occurred on October 23, 2004, are shown below. Table 5.8 shows the maximum acceleration values observed by each seismograph. Figure 5.21 shows the acceleration waveform of each seismograph, and Figure 5.22 shows the load-displacement curve of the brace-type oil damper on the observation floor.

Table 5.8 Acceleration at the measurement point

	Building			Ground	
Location	NS	EW	Location	NS	EW
RFL	21.5	19.7			
5FL	18.6	19.5			
1FL	9.2	9.9	GL	9.9	10.1
B3FL	8.1	9.1	GL-15	8.0	8.9
			GL-50	5.6	5.6

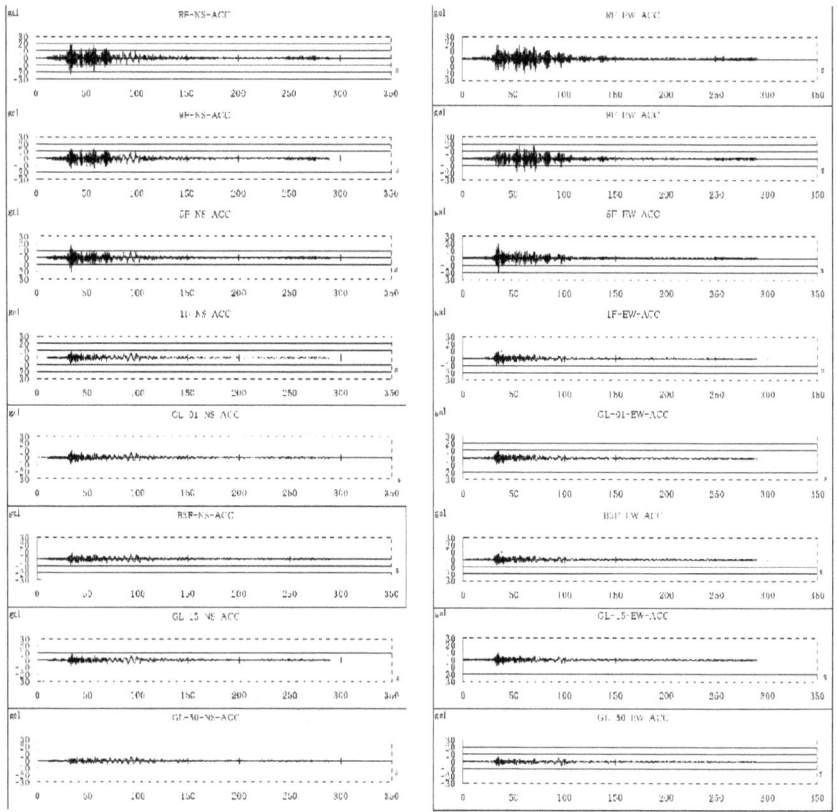

Figure 5.21 Input acceleration waveform.

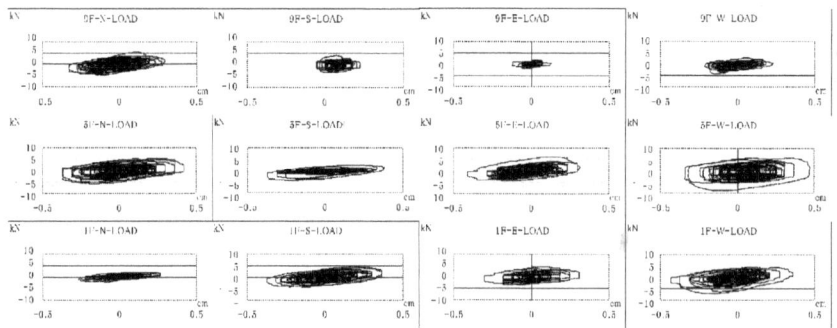

Figure 5.22 Displacement curve.

Every brace-type oil damper draws a beautiful loop, and it was confirmed that this vibration-damping device has a damping effect even though it has a small amplitude due to energy absorption.

5.8 SUMMARY

We introduced an example of installing a vibration-damping device that is integrated with the design in an actual building for the purpose of improving the habitability of the environmental vibration level of the building and the structural safety of the earthquake level.

In addition, the vibration-damping effect of this building was confirmed by a shaker experiment just before the completion, a human-powered vibration experiment two years after the completion, and long-term observations of earthquakes and winds that have continued after the completion. The main findings obtained are as follows.

(1) It was confirmed that a highly damped building can be realized by designing a brace-type oil damper to be attached to the actual building.
(2) It was confirmed that the attachment/detachment time of the joint pin was as short as about 5 minutes at one location even during construction, and that it could be attached/detached by a general steel frame construction operator.
(3) By comparing both the case where the brace-type oil damper is installed and the case where the brace-type oil damper is not installed in the experiment by the vibrator experiment just before the completion, even if the brace-type oil damper is installed, the natural period is reached. It was confirmed that there was no large fluctuation, the maximum displacement of the resonance curve was about 1/3, and the vibration-damping effect was exhibited.
(4) It was confirmed by a human-powered vibration experiment two years after the completion that the equivalent damping h = 0.04 to 0.05 was exhibited even at a very small amplitude level.
(5) In October 2004, the seismic observation results of the Chuetsu Earthquake were organized, and an elliptical loop peculiar to the viscous damper was drawn in the damping force-displacement relationship of any brace-type oil damper and this vibration suppression. It was confirmed by the device that the vibration-damping effect was exhibited by energy absorption even though the amplitude was small.

REFERENCES

1. Toshiko Okuzono, Osamu Takahashi, Masayuki Ninomiya: "Structural planning institute building; vibration control structure by directly attaching oil damper", *Building Technology*, pp.12-1–12-6, 1999.3.

2. Masayuki Ninomiya, Tomio Okabe, Toshiko Okuzono, Osamu Takahashi, Yuto Usami: "Development of seismic control structure using oil damper (Part 5. Long-term observation results)", *Architectural Institute of Japan Conference Academic Lecture Abstracts (Tohoku)*, pp.939–940, 2000.9.

3. Osamu Takahashi, Tomio Okabe, Toshifumi Okuzono, Masayuki Ninomiya: "Response control structure with oil damper bracing system", *EASEC7*, pp.871–876, 1999.8.

4. Osamu Takahashi, Tomio Okabe, Toshiko Okuzono, Masayuki Ninomiya: "Development of seismic control structure using oil damper (3. Application example to office building)", *Architectural Institute of Japan Conference Academic Lecture Summary (Kyushu)*, pp.937–938, 1998.9.

5. Osamu Takahashi, Tomio Okabe, Toshiko Okuzono, Masayuki Ninomiya: "Development of seismic control structure using oil damper (Part 4. vibration test)", *Architectural Institute of Japan Conference Academic Lecture Summary (China)*, pp.1049–1050, 1999.9.

6. Osamu Takahashi, Tomio Okabe, Toshiko Okuzono, Hitoshi Nakamura: "Development of seismic control structure using oil damper (6. Human-powered vibration experiment)", *Architectural Institute of Japan Conference Academic Lecture Summary (Hokuriku)*, pp.701–702, 2002.8.

7. Osamu Takahashi, Toshifumi Okuzono, Yukimori Yanagawa: "1061 response control structure with oil damper bracing system –Volume 2 vibration test and long term observation", *EASEC8*, pp.15-28–15-30, 2001.12.

Chapter 6

Analytical model and verification of building oil damper under small amplitude

6.1 INTRODUCTION

Oil dampers are often used in damping buildings for the purpose of reducing the response during an earthquake because of the advantages such as high efficiency of damping capacity compared to the size of the device and low influence of damping characteristics depending on the environment.[1]

In addition, since the building oil damper mechanically controls its damping characteristics, it is easy to construct an analysis model and simulate it by numerical calculation, and the calculation result is comparably precise at an amplitude level of 0.1 cm or more.[2] Therefore, it can be easily used if the structural designer understands the basic damping characteristics of the oil damper as a vibration-damping member and its applicable range.

On the other hand, in recent years, it has been expected that the damping effect of the oil damper will be exhibited even for relatively small disturbances such as daily wind response and traffic vibration. Therefore, research on the behavior of the oil damper as a vibration-damping member in the small amplitude region has also been studied.[3] However, also in the mechanical field, the oil damper is generally used for the purpose of reducing vibration in a region where the amplitude is large since anti-vibration rubber is often used in the small amplitude region. So, there is no research that proposes the modeling of oil dampers for the small amplitude. In addition, even in the field of building structures targeting small disturbances, there is research that concludes that the damping force-displacement relationship of the oil damper as a vibration-damping member has the effect of sliding resistance due to packing friction at the time of small amplitude.[4, 5, 6] The situation has not reached the proposal of quantitative modeling.

In the small amplitude region, it is conceivable that multiple factors such as (1) the relatively large sliding resistance due to the friction of the packing placed inside the oil damper to prevent oil leakage, (2) the decrease of the compression rigidity of the oil due to the internal air bubbles, and (3) the influence of the initial operation of the control valve, will become apparent, and their damping characteristics will differ from the basic characteristics.

DOI: 10.1201/9781003290261-7

In this chapter, we will discuss two types of building oil dampers (500 and 250 kN), both rod-type building oil dampers as shown in Figure 6.1, from a mechanical point of view in the small amplitude region. We propose a mechanical model and report on the accuracy of the analysis results.

6.2 BASIC MECHANICAL PROPERTIES OF BUILDING OIL DAMPERS

6.2.1 Basic principle of building oil damper

The basic principle of a general oil damper is a piston structure as shown in Figure 6.2. When the piston and the piston rod are pushed into the inside of the cylinder filled with oil by disturbance, the internal pressure rises and the oil flows, and a damping resistance force depending on the pushing speed is generated.

When the piston and piston rod inside of the oil-filled cylinder (φ: D (cm)) are pushed at a disturbance speed: u_d (cm/s), they flow at a flow rate: Q (cm^3/s) and internal pressure: P (N/cm^2) is generated by the viscous force and the flow resistance force.

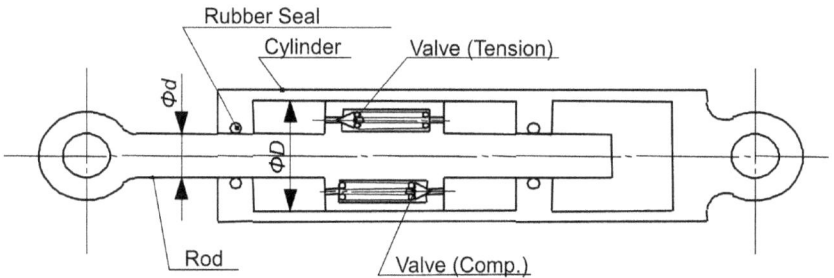

Figure 6.1 Basic structure of building oil damper.

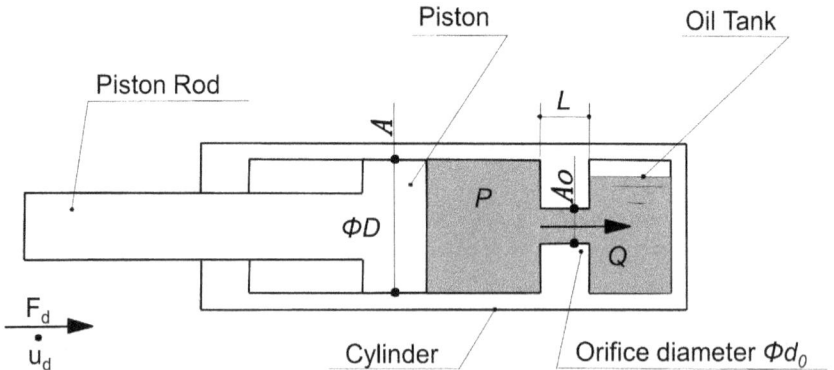

Figure 6.2 Basic principle of oil damper.

$$F_d = \left[8 \cdot \pi \cdot v \cdot L + \frac{A \cdot \dot{u}_d}{2 \cdot C_D{}^2} \right] \cdot \left(\frac{\rho \cdot A^2}{A_0{}^2} \cdot \dot{u}_d \right) \tag{6.1}$$

This internal pressure acts on the piston to form a resistance force F_d (N), and this force is called a damping resistance force and can be expressed by Eq. (6.1).

Here, A: Piston pressure receiving area ($\frac{\pi}{4} \cdot \varphi \cdot D^2$) (cm²)

A_0: Orifice area ($\frac{\pi}{4} \cdot \varphi \cdot d_0{}^2$) (cm²)

v: Fluid kinematic viscosity coefficient (cm²/sec)
ρ: Oil density (kg/m³)
L: Orifice length (cm)
C_D: Flow coefficient
u_d: Piston speed (cm/sec.)

The first term in brackets in Eq. (6.1) indicates the resistance generated by viscosity, and the second term indicates the resistance generated by turbulence. The kinematic viscosity coefficient of the first term: (cm²/sec.) is large, and the damper that dissipates vibration energy as heat becomes highly dependent on temperature. In the double-rod-type building oil damper targeted here, the Reynolds number inside the damper is about 2500 to 4000 by using hydraulic oil with low viscosity, and the orifice length: L is made small to make it large. The structure is such that the effect of viscosity in item 1 in parentheses can be ignored. The damping force characteristic equation of the structure in which the viscous resistance force can be ignored is shown in Eq. (6.2).

$$F_d = \left(\frac{\rho \cdot A^3}{2 \cdot C_D{}^2 \cdot A_0{}^2} \cdot \dot{u}_d \right) \tag{6.2}$$

Equation (6.2) has a characteristic proportional to the square of the velocity, and generally represents the generated resistance of a simple piston structure.

6.2.2 Basic characteristics of building oil damper

The damping resistance in the low-speed region becomes small since the characteristic that the damping resistance is proportional to the square of the speed. In the building oil damper, expecting high damping resistance even in the low-speed region, the damping resistance is often proportional to the first power of the velocity as shown in Figure 6.3. In addition, many construction oil dampers that exhibit a large damping resistance are equipped with a relief valve and have a fail-safe mechanism that suppresses the increase in load when a certain load is applied as shown in Figure 6.4.

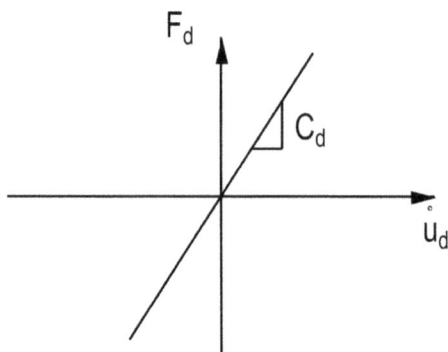

Figure 6.3 Basic characteristics (without relief valve).

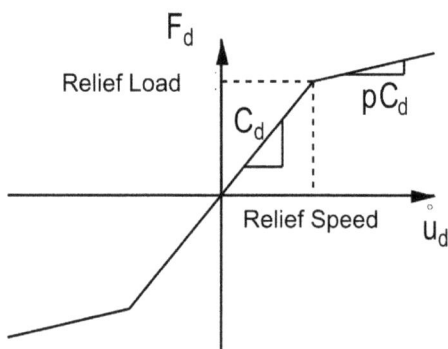

Figure 6.4 Basic characteristics (with relief valve).

A detailed diagram of the pressure regulating valve is shown in Figure 6.5. The pressure regulating valve has a mechanism proportional to the first power of the speed by adjusting the opening area of the valve port diameter of the spring. In addition, a constant orifice is provided to regulate the oil flow in the small speed range before the valve opens. Figure 6.6 shows a schematic diagram of the hydraulic circuit.

As can be seen from the schematic diagram of the hydraulic circuit, an initial load is applied to the spring of the pressure regulating valve, and the mechanism is such that it does not open unless it exceeds a certain load, so oil flows from the constant orifice before the valve opens. The damping characteristic after the valve is opened is controlled by a pressure-regulating valve at low speed and by operating a release valve at high speed.

In order to make the damping resistance force proportional to the first power of the speed, the pressure regulating valve keeps the internal pressure and the balanced state, by pressing the conical valve with a spring against the valve port surface as shown in Figure 6.5. When the internal pressure

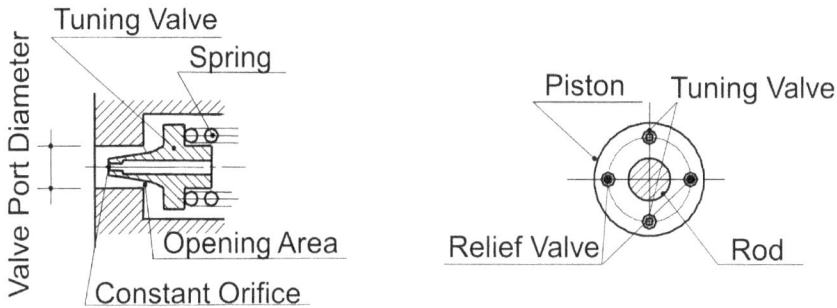

Figure 6.5 Detail of pressure tuning valve.

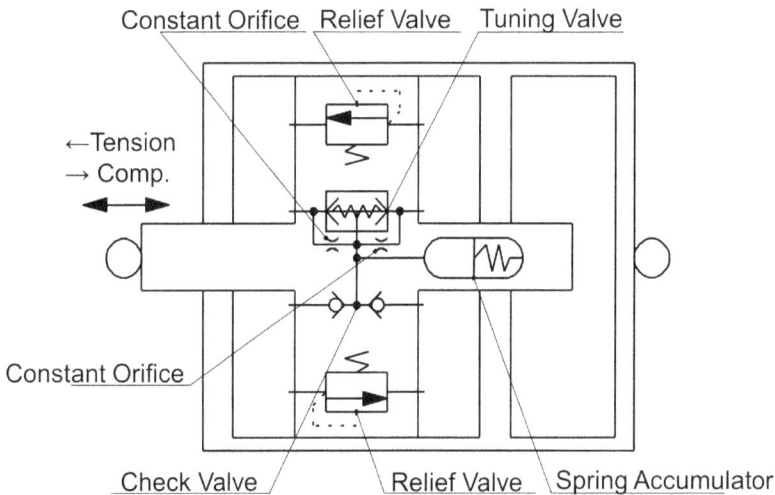

Figure 6.6 Schematic diagram of hydraulic circuit.

becomes larger than the spring force, the valve opens and a gap is created between the valve port diameter and the conical valve, and oil flows out. The relationship between the lift amount X(cm) and the opening area $A(x)$ (cm^2) at this time is given by Eq. (6.3).

$$A_{(X)} = \lambda \cdot \sqrt{X} \qquad\qquad (6.3)$$

Here, λ is a proportional constant, and the shape of the pressure regulating valve is set so that it becomes (= 0.01) in the target oil damper.

On the other hand, the balance between the pressure acting on the pressure regulating valve: P (N/cm^2) and the spring force is represented by Eq. (6.4), assuming that the pressure receiving area of the valve port is A_v (cm^2) and the spring constant is K_v (N/cm).

$$P \cdot A_v = K_v \cdot X \tag{6.4}$$

The relationship between damping resistance: F_d (N) and pressure: P (N/cm^2) is given by Eq. (6.5).

$$P = \frac{F_d}{A} \tag{6.5}$$

By substituting Eqs. (6.4) and (6.5) into Eq. (6.3) and replacing the orifice area of Eq. (6.2): A_0 (cm^2) with Eq. (6.3), Eq. (6.6) can be derived.

$$F_d = \sqrt{\frac{\rho \cdot K_v \cdot A^4}{2 \cdot A_v \cdot \lambda^2 \cdot C_D{}^2}} \cdot \dot{u}_d \tag{6.6}$$

Here,

$$F_d = C_d \cdot \dot{u}_d \tag{6.7}$$

Then, Eq. (6.6) becomes as follows.

$$C_d = \sqrt{\frac{\rho \cdot K_v \cdot A^4}{2 \cdot A_v \cdot \lambda^2 \cdot C_D{}^2}} \tag{6.8}$$

Utilizing these mechanical controls, a device having a damping resistance that is proportional to the first power of the velocity is generally used as the damping characteristic of the building oil damper. C_d is called the viscosity damping coefficient and is a set value set by the structural designer when designing a building as a characteristic proportional to the first power of the velocity.

The flow coefficient of the building oil damper: C_d ($=\sqrt{0.5}$) is set based on the shape of the pressure regulating valve, and the oil density is ρ ($= 0.00088$ (kg/m^3)).

Figure 6.7 shows the damping force-displacement relationship during sinusoidal vibration of a building oil damper with such characteristics. The one without a relief valve shows an inclined ellipse, and the one with a relief valve shows a damping force-displacement relationship close to a parallelogram with the top and bottom of the ellipse cut. The damping force-displacement relationship of the building oil damper has a slope mainly due to the existence of internal rigidity due to the compressive rigidity of the oil. The damping force-displacement relationship, which indicates the energy absorption amount of the oil damper, shows a thin elliptical shape with the damping force-displacement-related loop shape when the rigidity is low. [7] Depending on the evaluation of the material rigidity of the members, damper cylinders, rods, etc., the damping force-displacement relationship loop shape is greatly affected. Since this internal rigidity K_d is smaller than

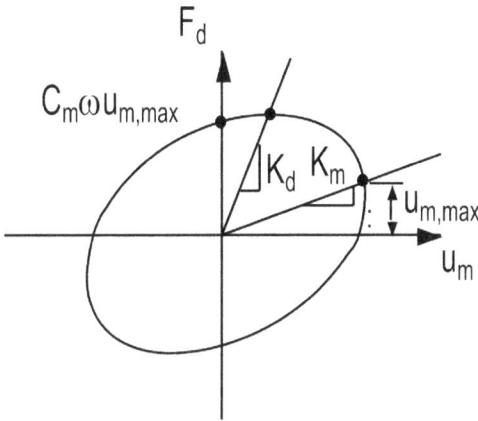

Figure 6.7(a) Damping force-displacement relationship (under sine wave excitation).

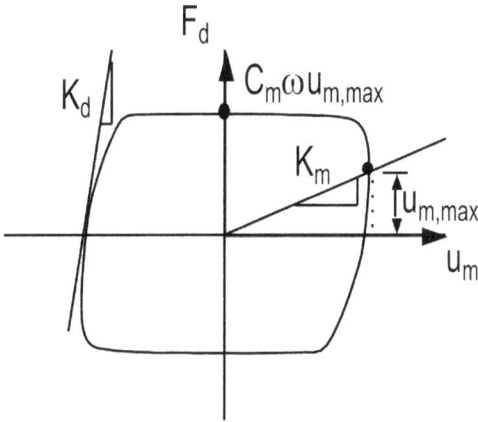

Figure 6.7(b) Damping force-displacement relationship (under sine wave excitation).

the material rigidity of damper mounting members made of steel, cylinders, rods, etc. of dampers, it has a large effect on the damping force-displacement relationship. When performing an analysis, it is necessary to properly evaluate it in the analysis model.

6.2.3 Basic analysis model of building oil damper

In order to consider the above-mentioned basic characteristics in the analysis model of the building oil damper, the nonlinear Maxwell model in which the mounting members are arranged in series in the damper part having the dash pot with the viscosity damping coefficient C_d and the linear spring with the internal rigidity K_d, is commonly used as shown in Figure 6.8.

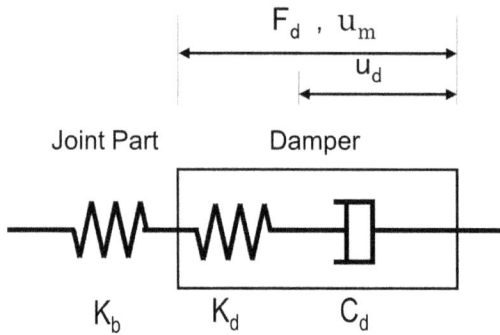

$$F_d \; , \; u_m$$
$$u_d$$

Joint Part   Damper

$$K_b \qquad K_d \qquad C_d$$

Figure 6.8 Nonlinear Maxwell model.

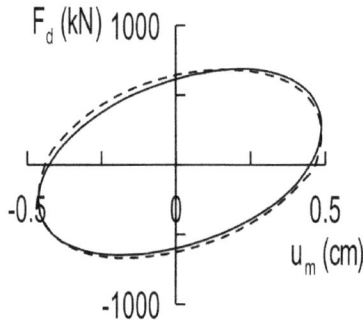

F_d (kN) 1000

-0.5   0   0.5
u_m (cm)

-1000

Figure 6.9(a) Analysis result example of history curve of nonlinear Maxwell model (max. amp.: 5 mm).

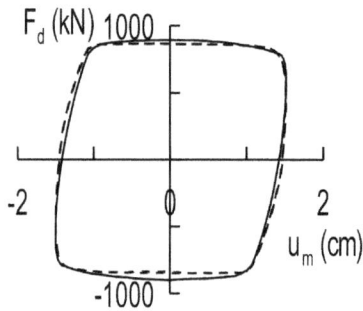

F_d (kN) 1000

-2   0   2
u_m (cm)

-1000

Figure 6.9(b) Analysis result example of history curve of nonlinear Maxwell model (max. amp.: 15 mm).

The time history response analysis results of the building oil damper using the nonlinear Maxwell model well correspond to the experimental results as shown in Figure 6.9 in the amplitude range of about 0.1 cm or more. [2] On the other hand, the analysis results in the range of amplitude of about 0.1 cm or less are inaccurate with the experimental results, and the above-mentioned nonlinear Maxwell model has been reported not to be suitable as a mechanical model of the building oil damper in the small amplitude range of about 0.1 cm or less.[2, 3]

6.3 SMALL AMPLITUDE MODEL OF BUILDING OIL DAMPER

6.3.1 Damping force–velocity relationship

An initial load is applied to the pressure-regulating valve provided in the oil flow path, the oil does not operate within a certain range, and the damping force is generated by the oil passing through the constant orifice provided in series with the pressure regulating valve. In this case, the damping force below a certain velocity is not proportional to the velocity but shows the damping characteristic proportional to the square of the velocity as shown in Figure 6.10. Therefore, the damping force of the oil damper is expressed by the following equation.

$$F_d = C_x \cdot u_d^2 \qquad (u_d < U_x) \tag{6.8.1}$$

$$F_d = C_d \cdot u_d \qquad \left(u_d \geqq U_x\right) \tag{6.8.2}$$

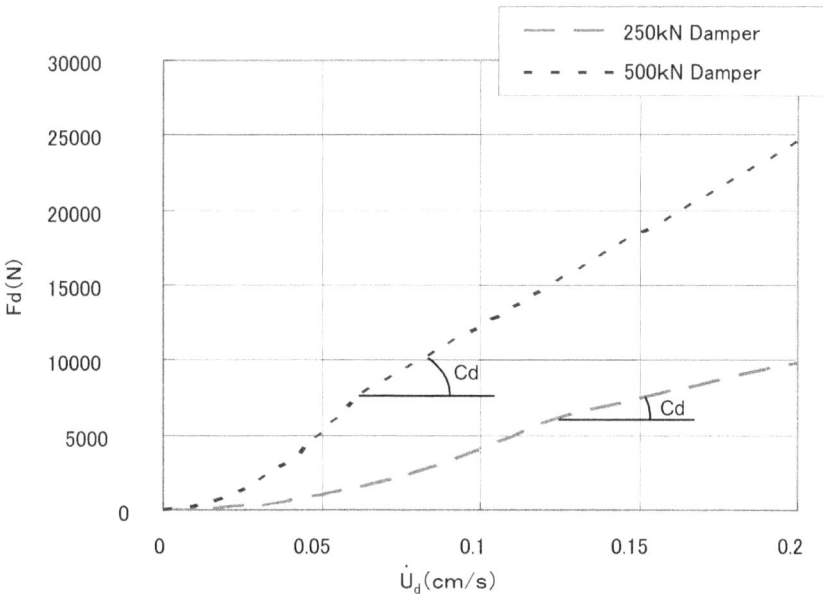

Figure 6.10 Damping force-velocity relationship (under small amplitude).

Here,

C_x: Proportional coefficient of velocity square proportional range

U_x: Switching speed of proportional characteristics

The switching speed: U_x can be obtained by the following equation from the intersection of the proportional equation of the speed and speed squared.

$$U_x = \frac{C_d}{C_x} \tag{6.9}$$

In addition, the coefficient of the proportional range of velocity squared: C_x is a numerical value determined by the volume of oil and the density of oil and is constant depending on the shape of the damper.

6.3.2 Compressive stiffness of oil

Since the air mixed in the oil exists as bubbles and cannot be completely removed, the compression rigidity of the oil is lowered due to the influence of the bubbles in the small amplitude region where the internal pressure is small. The apparent bulk modulus of oil containing bubbles is expressed by the following equation, assuming that the compression of bubbles is adiabatic compression.[8]

$$K_a = \frac{1 + \left(\dfrac{x_0}{1 - x_0} \right) \cdot \left(\dfrac{P_0}{P} \right)^{\frac{2}{\kappa}}}{1 + \left(\dfrac{x_0}{1 - x_0} \right) \cdot \left(\dfrac{P_0}{P} \right)^{\frac{1}{\kappa}} \cdot \left(\dfrac{K_0}{\kappa \cdot P} \right)} \cdot F_{ij} \tag{6.10}$$

Here,

K_0: Bulk modulus of oil (N/cm²)

K_a: Bulk modulus of oil containing bubbles (N/cm²)

x_0: Air bubble mixing rate (value determined by oil = 0.005)

P_0: Atmospheric pressure (N/cm²)

P: Internal pressure (kN/cm²)

κ: Adiabatic index (air adiabatic index = 1.4)[9]

The bulk modulus of oil: K_0 is expressed by the following equation.[10]

$$K_0 = V_c \cdot \frac{dP}{dV} \tag{6.11}$$

Here,

V_c: Initial volume of oil (= pressure receiving area × stroke)

dP: Pressure change

dV: Change in oil volume due to compression

The bulk modulus of oil is generally 1.4 to 1.9 (MPa). This elastic modulus is expressed as a spring arranged in series with the dashpot in the Maxwell model shown in Figure 6.8 as a damper. If oil is an elastic body, the rate of change is

$$dF_d = dP \cdot A \tag{6.12}$$

The rate of change in displacement is

$$du_d = \frac{dV}{A} \tag{6.13}$$

The compression rigidity of oil K_{d0} when the air mixing ratio is not taken into consideration is ...

$$K_{d0} = \frac{dF_d}{du_d} \tag{6.14}$$

And K_0 is expressed by substituting Eqs. (6.12) and (6.13) into Eq. (6.14) and using Eq. (6.11).

$$K_{d0} = \frac{K_0 \cdot A^2}{V_c} \tag{6.15}$$

When considering the air mixing ratio, K_0 in Eq. (6.15) is K_a, and the following equation is obtained. Internal stiffness of oil containing air bubbles: K_{da} shows the relationship with the internal pressure as shown in Figure 6.11.

$$K_{da} = \frac{K_a \cdot A^2}{V_c} \tag{6.16}$$

The bubble mixing rate x_0 is expressed by the following equation and indicates the ratio of the volume of bubbles contained in the oil under atmospheric pressure. The mixing rate of air bubbles is also affected by the lubrication method, but because of measuring before lubrication with an air volume measuring device using the method obtained by reducing the pressure in the air chamber, the oil used in the oil damper of this study is, we have confirmed, 0.5%.

$$x_0 = \frac{V_{a0}}{V_{f0} + V_{a0}} \tag{6.17}$$

V_{a0}: Volume of bubbles contained in oil

V_{f0}: Volume of oil only

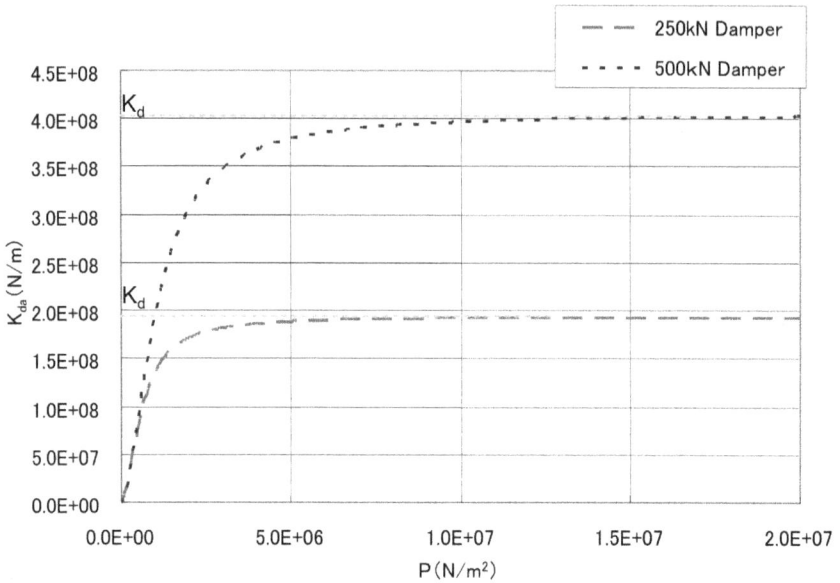

Figure 6.11 Pressure dependence of oil compression stiffness.

6.3.3 Sliding resistance

Since the packing used for the sliding portion of the piston rod has resistance due to friction, it is considered that this effect cannot be ignored in the small amplitude region where the generated resistance force is small.

In addition, in the type of packing used for the oil damper targeted this time, this sliding resistance force: F_f increases depending on the internal pressure inside the cylinder.[11] Figure 6.12 shows the Pressure dependence of sliding resistance force Ff.

6.3.4 Analytical model

To express the mechanical properties of the compressive rigidity of oil and the sliding resistance due to packing considering the mixing rate of bubbles, the mechanical model shown in Figure 6.13 is proposed here. In this model, the oil pressure considers the air bubble mixing rate.

A Maxwell element consisting of a spring with compression rigidity and a dashpot whose speed dependence characteristics change, and a friction element expressing the sliding resistance force by packing are arranged in parallel, and the damper part and the spring with mounting member rigidity are arranged in series.

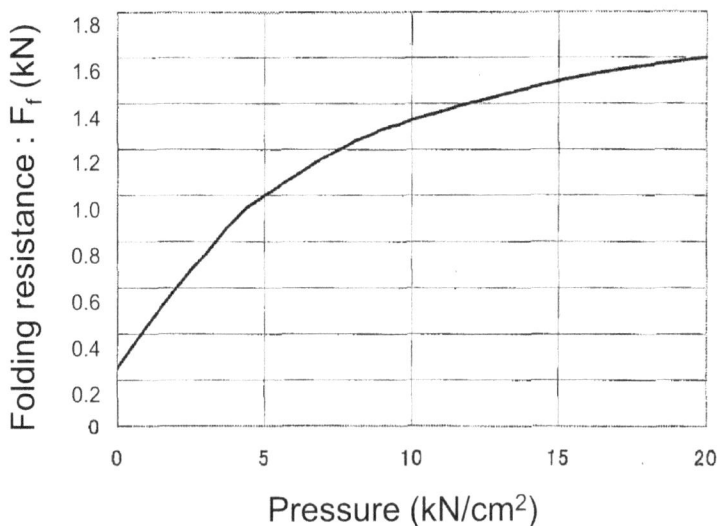

Figure 6.12 Pressure dependence of sliding resistance force.

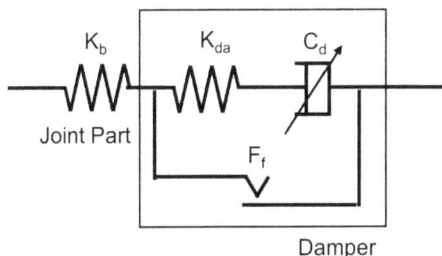

Figure 6.13 Dynamics model under small amplitude.

6.4 APPLICATION EXAMPLE OF SMALL AMPLITUDE MODEL (COMPARISON WITH SMALL AMPLITUDE EXCITATION EXPERIMENT)

6.4.1 Experiment outline

A sine wave vibration experiment with a small amplitude has been conducted for a building oil damper with a damper capacity of 500 kN.[3] As shown in Figure 6.14, the test equipment consists of a shaking table and a weight mounted on it and is a one-mass point system to which an oil damper is directly attached. The vibration frequency is set to 0.25 to 3.0 Hz as shown in Figure 6.15, which is considerably smaller than the resonance frequency (10.8 Hz) of the system. From that condition, the displacement

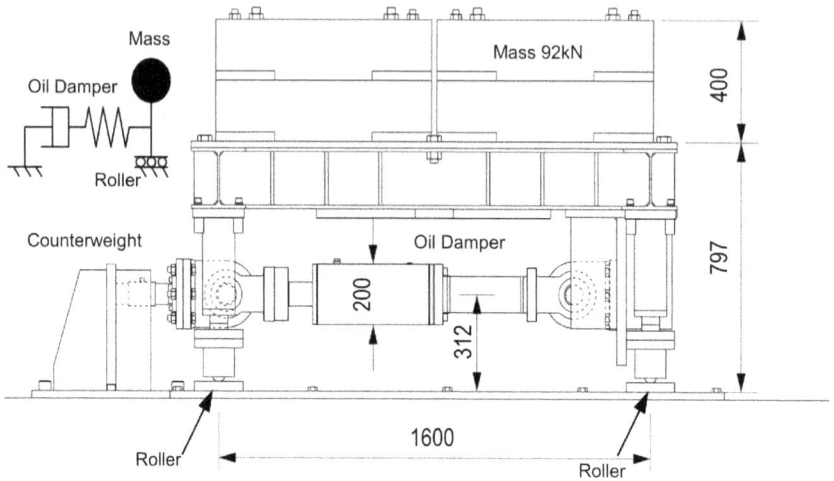

Figure 6.14 Test equipment and oil damper.

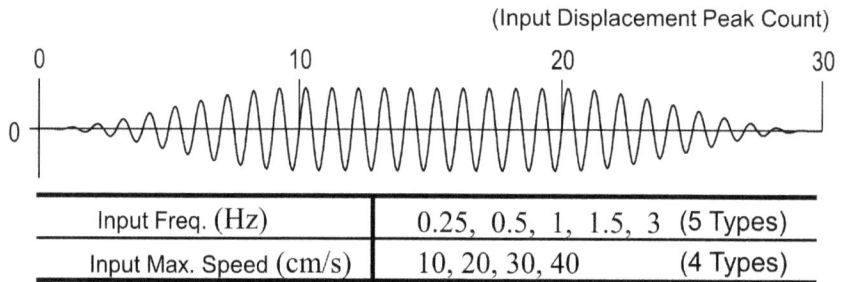

(Input Displacement Peak Count)		

Input Freq. (Hz)	0.25, 0.5, 1, 1.5, 3 (5 Types)
Input Max. Speed (cm/s)	10, 20, 30, 40 (4 Types)

Figure 6.15 Excitation plan.

of the damper is in small amplitude range and the shaking table can work stably by large amplitude sine-wave.

6.4.2 Setting of analysis parameters

The parameters used in the analysis are shown in Table 6.1. Compressive stiffness of oil considering the mixing rate of bubbles: K_{da} depends on the pressure as described above, but here, for the sake of simplification of calculation, the average compressive stiffness until the maximum pressure is obtained is adopted and kept constant. As shown in Figure 6.16, the average compressive stiffness: K_{da_ave} is the value obtained by dividing the area ΔPK of the shaded area obtained at the maximum pressure: P_{max} by P_{max}. In addition, the air bubble mixing rate: x_0 is affected by the lubrication method, but the value of 0.5% of the result measured before lubrication with an air volume measuring device was adopted. Table 6.2 shows the values of oil compression stiffness used in the analysis.

Table 6.1 Analysis parameters

C_d	1.23E+05 (N·s/cm)
C_x	2.02E+06 (N·s²/cm²)
U_x	6.01E-02 (cm/s)
K_0	1.63E+05 (N/cm²)
χ_0	0.005
P_0	9.81 (N/cm²)
A	1.5E+01 (cm²)
V_c	9.05E+02 (cm³)

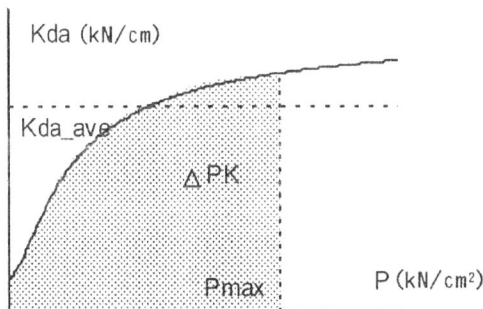

Figure 6.16 Calculation method of average of oil compressive stiffness.

Table 6.2 List of oil compression rigidity

Device max. speed (cm/s)	10	20	30	40
f = 0.25 Hz	0.02	0.02	0.02	0.6
f = 0.5 Hz	0.42	0.58	0.75	0.97
f = 1.0 Hz	0.66	0.92	1.22	1.45
f = 1.5 Hz	0.75	1.22	1.54	1.79
f = 3.0 Hz	1.31	1.9	2.14	2.29

(Unit: ×10⁵ N/cm)

6.4.3 Analysis result

Figure 6.17 shows a comparison of the historical curves of the experimental results and the analysis results for each input frequency. It can be said that the analysis results generally correspond to the experimental results.

However, there are some places where the error is clear, and one of the causes is considered to be the turbulence of the oil flow near the switching point of the velocity proportional characteristics.

Table 6.3 shows the maximum response speed converted from the maximum displacement of the shaking table for each frequency.

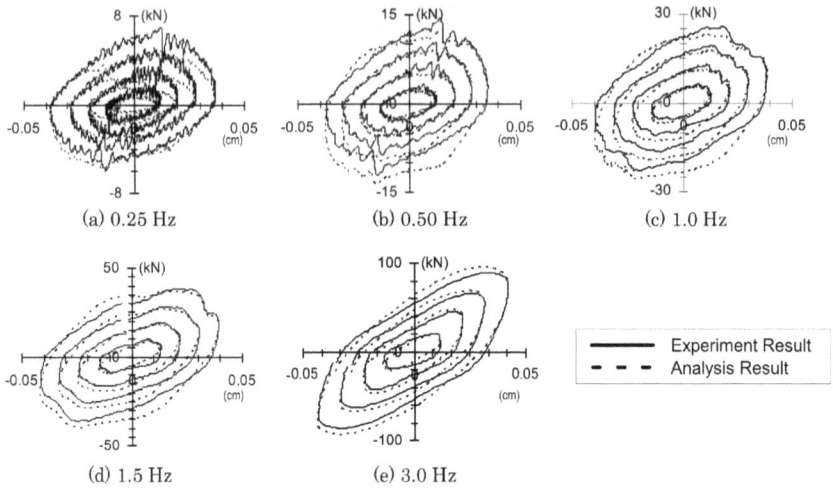

Figure 6.17 Damping force-displacement relationship (experiments and analysis).

Table 6.3 Maximum response velocity (calculated from maximum displacement)

Device max. speed (cm/s)	10	20	30	40
$f = 0.25$ Hz	0.2	0.32	0.45	0.6
$f = 0.5$ Hz	0.42	0.69	0.94	1.18
$f = 1.0$ Hz	0.86	1.41	1.95	2.45
$f = 1.5$ Hz	1.38	2.27	3.09	3.93
$f = 3.0$ Hz	3.1	5.11	7.49	9.75

(Unit: $\times 10^{-2}$ N/cm)

6.5 EXAMPLE OF APPLICATION OF THE SMALL AMPLITUDE MODEL (COMPARISON WITH BUILDING FREE VIBRATION EXPERIMENT)

6.5.1 Experiment outline

An excitation experiment was conducted on a steel-framed five-story office building (five floors above ground, total floor area 1,765 m²) under construction using a brace-type oil damper as a vibration-damping device.[4, 5, 6] Figure 6.18 shows the shape of the building and the layout of the oscillator. The vibration-damping device is equipped with a brace-type oil damper[7] arranged in series with the brace.

The exciter used in the experiment was fixed via an anchor attached to a steel beam in the center of the roof of the building. In the experiment, after operating the exciter, the mass was fixed to generate free vibration of

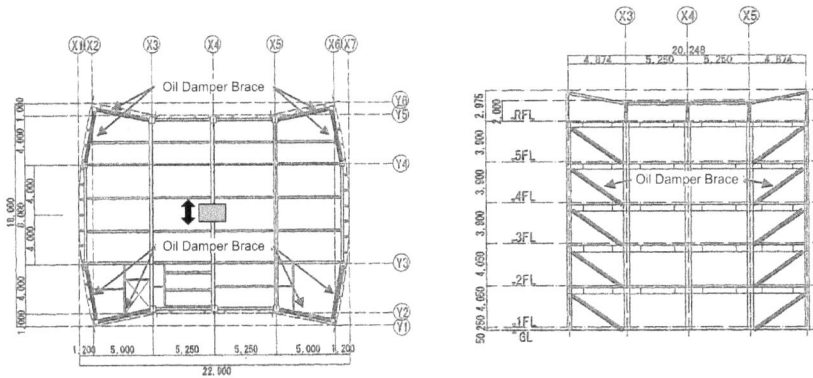

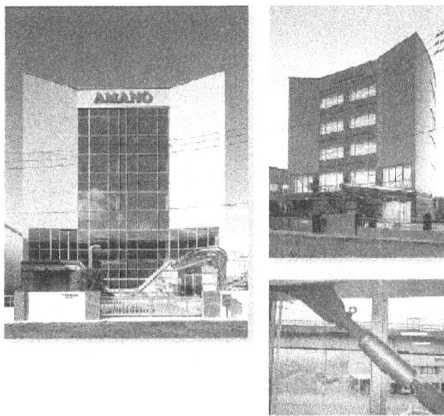

Figure 6.18 Building and exciter installation.

the building, and the damper stroke by the displacement meter (differential transformer type) attached to the brace-type oil damper and the shaft strain of the brace were measured.

6.5.2 Setting of analysis parameters

Tables 6.4 to 6.6 show a list of each parameter used in the analysis.

6.5.3 Analysis results

The comparison between the experimental results and the analysis results is shown in Figures 6.19 and 6.20. The analysis results are almost in agreement with the experimental results, indicating that the above-mentioned proposed model is effective. For reference, Figures 6.19 and 6.21 also show the case where the sliding resistance is not considered, but the analysis

Table 6.4 Building overview

location	Okayama, Okayama
Site area	1248.0 (m²)
Total floor area	1785.2 (m²)
Number of stories	5 floors above ground
Structural format	Steel frame
Damping device	Brace-type oil damper (40 oil dampers with a max. damping force of 250 kN type are used.)

Table 6.5 Outline of the exciter

Excite method	Slide mass type
Excite force direction	Horizontal one direction
Max. excite force	9.8 (kN) (both amplitudes)
Excite waveform	sine wave
Body weight	16 (kN)
Body dimensions (LxDxH)	2.499 x 1.35 x 1.017 (m)
Frequency range	0.1 ~ 10.0 (Hz)
Frequency resolution	0.001 (Hz)

Table 6.6 Analysis parameters

C_d	4.90E+04 (N·s/cm)
C_x	3.96E+05 (N·s²/cm²)
U_x	1.24E-01 (cm/s)
K_0	1.63E+05 (N/cm²)
x_0	0.005
P_0	9.81 (N/cm²)
A	7.13E+01 (cm²)
V_c	4.28E+02 (cm³)
K_{da}	2.50E+05 (N/cm)

result by the model that does not consider the sliding resistance is the result that the damping force is smaller than the experimental result. The result is that there is no agreement between the damping force and the displacement.

Comparing the damping force-displacement relationship, the experimental results and the analysis results are almost the same in the model considering the sliding resistance force. The maximum speed is about 0.06 (cm/sec), which is a region below the value of the switching speed U_x of the proportional characteristic, but it can be confirmed that the sliding resistance has a large effect on the damping force-displacement relationship.

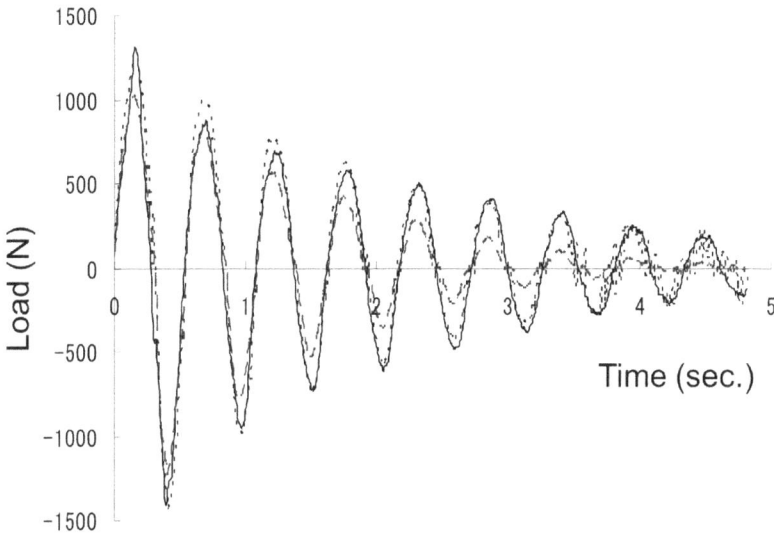

Figure 6.19 Damper-load time history.

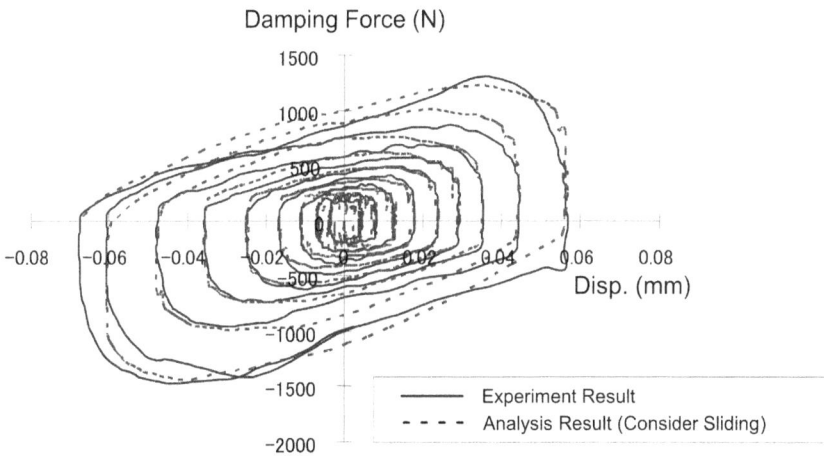

Figure 6.20 Damping force-displacement relationship (with sliding consideration).

6.6 SUMMARY

We constructed and proposed an analytical model that expresses the behavior of a building oil damper in the small amplitude region based on its internal mechanism. By appropriately considering the effects of damping force-velocity relationship, oil compression rigidity, and packing sliding resistance force in the small amplitude region, it is confirmed that the

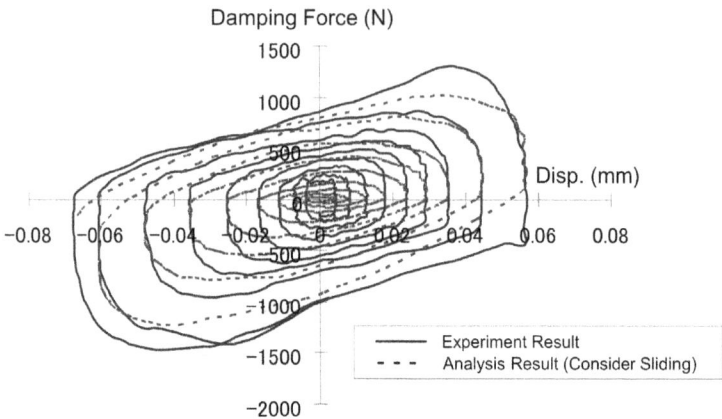

Figure 6.21 Damping force-displacement relationship (without sliding consideration).

effectiveness of this modeling for the past experimental results is also shown in the brace-type oil.

In the future, if the structural designer expects a damping effect against relatively small disturbances such as daily wind and traffic vibration, the sliding resistance inside the building oil damper will have a large effect in the small amplitude region. It is desirable to adopt a building oil damper for the building after clearly understanding that it is affected by the mixing rate of the internal air bubbles and that it behaves differently from the basic characteristics at the initial operation of the control valve.

REFERENCES

1. Toshiko Okuzono, Osamu Takahashi, Masayuki Ninomiya: "Structural planning research institute building; vibration control structure by directly attaching oil damper", *Steel Structure Technology*, pp.12.1–12.6, 1999.3.
2. Osamu Takahashi, Yohei Sekiguchi: "Analysis algorithm and subroutine of oil damper using Maxwell model", *Passive Vibration Damping Structure Symposium 2001*, 2001.12.
3. Yoji Oki, Kazuhiko Kasai, Osamu Takahashi: "Structural engineering papers on the performance of velocity-dependent dampers at small amplitudes", 50B, pp.601–610, 2004.3.
4. Atsushi Onaka, Mayuya Soda, Osamu Takahashi, Tomio Kanazawa, Koji Oka, Hidehiko Itaya: "Seismic control structure by direct oil damper attachment method (1. Application example to office building)", *Architectural Institute of Japan Conference Academic Lecture Summary (Tohoku)*, B-2 Volume, pp.933–934, 2000.9.

5. Hidehiko Itaya, Mayuya Soda, Osamu Takahashi, Tomio Kanazawa, Atsushi Onaka, Koji Oka: "Seismic control structure by direct oil damper attachment method (Part 2. Outline and results of vibration experiment)", *Architectural Institute of Japan Conference Academic Lecture Abstracts (Tohoku)*, B-2 Volume, pp.935–936, 2000.9.
6. Koji Oka, Mayuya Soda, Osamu Takahashi, Koumori Ryokawa, Atsuyoshi Onaka: "Seismic control structure by direct oil damper attachment method (3. Train vibration)", *Architectural Institute of Japan Conference Academic Lecture Summary (Tohoku)*, B-2 separate volume, pp.937–938, 2000.9.
7. Osamu Takahashi, Yasuo Rouki, Naofumi Igarari, Ikuhiro Matsuzaki, Takashi Fujita: "Study on damping characteristics and analytical model of brace type oil damper", *Architectural Institute of Japan Structural Papers*, No.594, pp.57–64, 2005.8.
8. Toshio Takenaka, Shozo Urata: *Oil Mechanics*, Yokendo, 1970.2.
9. *Machinery Practical Handbook*, Revised 5th Edition, Japan Society of Mechanical Engineers, 2004.3. https://www.amazon.com/Machine-practical-handbook-ISBN-4888982090/dp/4888982090
10. *Practical Hydraulic Pocket Book*, 2001 Edition, Japan Fluid Power Industry Association, 2001.5. http://www.jfpa.biz/op_02.html
11. *SPNS Type Packing Technical Data*, Nippon Oil Seal Industry Co., Ltd.

Chapter 7

Conclusion to Part I

7.1 SUMMARY OF THIS STUDY

In this study, we investigated the development and analysis model of a brace-type oil damper considering the damping characteristics and various dependencies. The findings obtained from each chapter and the research results are summarized below.

In Chapter 1, "Introduction," the social necessity of damping structures was described as the background of the research, and the damping members were organized and the superiority of oil dampers was focused on. The findings and problems obtained from previous studies on damping structures were organized, and the position and purpose of this paper was clarified.

In Chapter 2, "Development of an oil damper for buildings considering damping characteristics and various dependencies," the development of building oil dampers that can be attached in the brace shape in consideration of damping characteristics and various dependencies was discussed. Here, the structural outline and hydraulic circuit diagram of the building oil damper were shown, and the points developed by focusing on the new building oil damper were clarified. Regarding the unit experiment of the building oil damper, as an experiment for confirming various dependent characteristics, a temperature dependence confirmation experiment, a temperature rise confirmation experiment, a durability confirmation experiment, and a random wave response confirmation experiment were conducted. As a result, we obtained knowledge about the range in which stable performance is demonstrated after systematically organizing the dependent characteristics depending on the temperature, speed, and frequency.

In Chapter 3, "Analytical model of a single building oil damper," a nonlinear Maxwell model was presented as an analysis model of a building oil damper so that it can be attached to a brace shape, and a numerical calculation algorithm was proposed. Quantitative consistency was confirmed by comparison with various dependent characteristics conducted in Chapter 2, and it was found that the proposal for modeling and solving the nonlinear Maxwell model was correct. We also obtained knowledge about the

DOI: 10.1201/9781003290261-8

problems of the analysis model and analysis parameters in the range where the amplitude is small.

In Chapter 4, "Damping characteristics and analytical model of brace-type oil damper," in order to conduct an assessment of the damping characteristics and analysis model, a brace-type oil damper was manufactured with the actual length assuming that it would be installed in an actual building. Also, it was found that a full-scale dynamic vibration experiment or a full-scale frame dynamic vibration experiment can be performed using an oil damper, and the consistency can be examined and quantitatively evaluated by comparing the experimental results with the proposed analysis model. In addition, we confirmed the safety and workability at the actual size level and obtained the knowledge that there is a possibility of practical use.

In Chapter 5, "Evaluation of vibration-damping performance of actual building using brace-type oil damper and confirmation experiment," a structural example of a building using a brace-type oil damper was shown. Confirmation of the damping effect by the vibration-damping experiment, the human-powered vibration experiment, and the long-term observation was also shown. In the shaker experiment, we compared both with and without the brace-type oil damper in the experiment and found that the brace-type oil damper exerts a vibration-damping effect by absorbing energy from a minute area in small and medium-sized earthquakes.

In Chapter 6, "Analytical model and verification of building oil damper under small amplitude" it was necessary to propose and examine a more detailed analysis model to assess building oil damper under minute amplitude such as environmental vibration. We proposed the analysis model and clearly showed the quantitative consistency by comparing with the actual measurement results as an application example of the minute amplitude model. When proposing a more detailed analysis model, it was found that it is necessary to pay attention to the sliding resistance inside the building oil damper and the mixing rate of air bubbles inside the oil.

In Chapter 7, "Conclusion," this study was summarized, and the future potential of brace-type oil dampers was presented. As a future possibility of the brace-type oil damper, we proposed an improved oil damper with a characteristic restoring force characteristic by improving the internal mechanism.

As an appendix, an example of a skyscraper and a reinforced building is shown as a practical example in an actual building using a brace-type oil damper.

From the above studies, it can be said that the damping performance and analysis method of the brace-type oil damper newly developed in this paper are effective and practical. Since the various parameters set in the analysis are practical, it is possible that they will be widely used not only for new properties from mid-low to high-rise buildings but also for seismic retrofitting of existing reinforced concrete structures and detached houses.

7.2 THE FUTURE OF THE BRACE-TYPE OIL DAMPER

The oil damper used in the brace-type oil damper is a customized version of the existing oil damper for construction. The oil damper has been widely used mainly in the transportation industry since the 1945s as a compact and high-performance vibration control device because it efficiently converts the input vibration energy into heat energy and dissipates it to the outside air. Regarding the temperature characteristics, it has been confirmed that the effect on the response at −20°C to +80°C is almost negligible when the oil with a high viscosity index, as used in aircraft, is used.

In the case of a building oil damper, by improving the internal mechanism for the failure safety of the outer column beam member and providing a release valve, the release valve can be used for damping force above a predetermined value. The characteristic is that the opening damping force is controlled. In the future, as an application of these, if the yield strength of the outer peripheral beam member cannot be expected, it is conceivable to use an improved oil damper showing a history loop as shown in Figure 7.1.

The single-cut type has the advantage that the floor response acceleration can be suppressed because the damping force due to the oil damper at the time when the deformation is maximized in the building response is smaller than that of the conventional type.

Both cut types have the advantage that the damping force due to the oil damper at the time of large deformation is smaller than that of the conventional type in the building response so that the influence on the peripheral members can be suppressed.

Oil Damper Type	Q – δ	Feature
This study	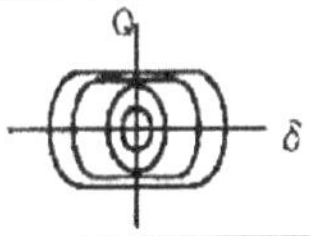	· Cut against load · Used as a fail-safe mechanism
One-way Active	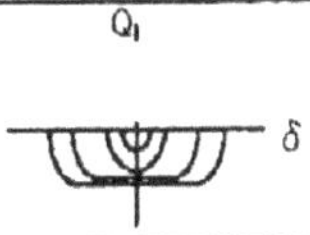	· Thin Damper Diameter · Adaptable to design space
One-way Cut	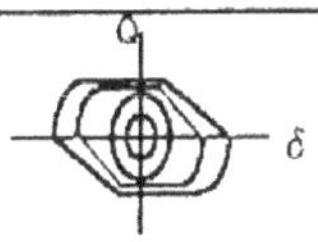	· Can reduce building response acceleration · Suitable for new buildings
Both-way Cut	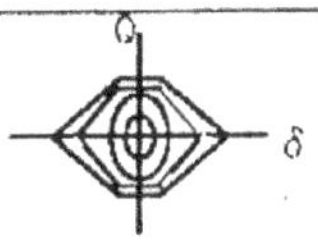	· Can reduce stresses in the damper's surrounding components · Suitable for reinforcements and detached houses.

Figure 7.1 Restoring force characteristics of improved oil damper.[1, 2]

Figures 7.2 and 7.3 show the outline of the principle of the improved oil damper.

In both the single-cut type and double-cut type, by processing a groove in the boundary between the outer cylinder and the piston where oil flows into

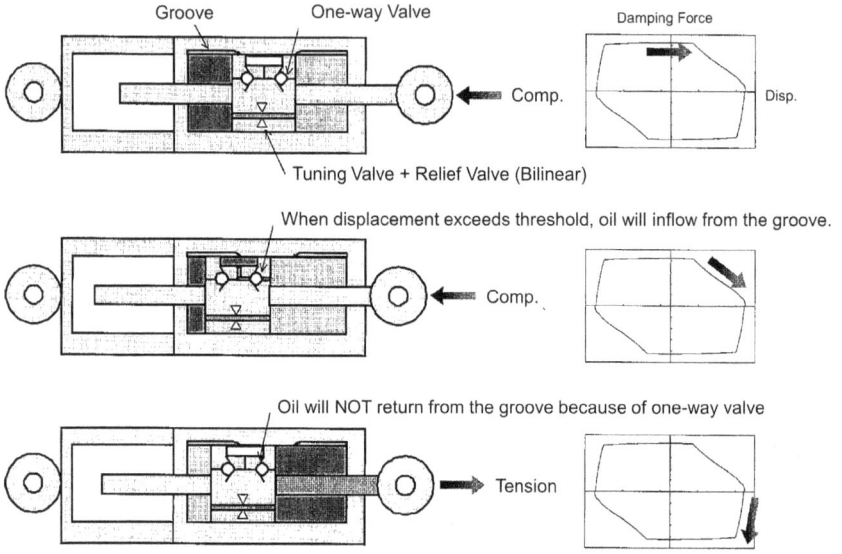

Figure 7.2 Principle of the improved oil damper (single-cut type).

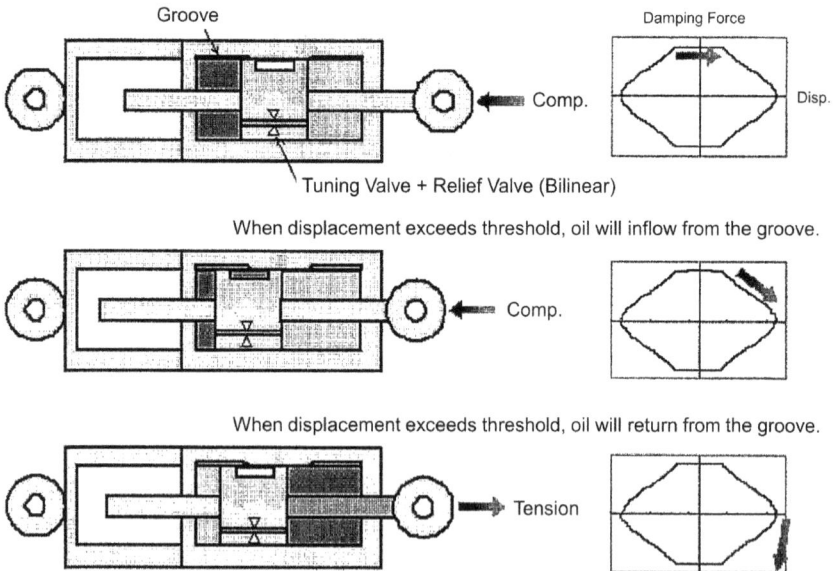

Figure 7.3 Principle of the improved oil damper (double-cut type).

the outer cylinder, oil exceeding a predetermined amount of deformation flows in from the groove and the damping force increases. It is a mechanism to suppress.

In both cut types, a one-way valve is provided to prevent oil from flowing in from the groove, and the mechanism is such that the damping force increases as in the conventional type.

In the future, it is expected that vibration control systems using these characteristic improved oil dampers will become widespread.

REFERENCES

1. Osamu Takahashi, Tomio Okabe, Toshiko Okuzono, Hitoshi Nakamura, Takashi Kawaso, Masanori Ogura: "Development of intelligent passive oil damper for building structures part 1 development overview", *Architectural Institute of Japan Conference Academic Lecture Summary*, B-2 separate volumes, pp.777, 2003.
2. Takashi Kawaso, Masanori Ogura, Osamu Takahashi, Tomio Okabe: "Development of intelligent passive oil dampers for building structures part 2 development of oil dampers", *Architectural Institute of Japan Conference Academic Lecture Abstracts*, B-2 Volume, pp.779, 2003.

Development of the oil damper stiffness for architectural vibration control and experimental research on structural characterization

Chapter 8

Preface

8.1 RESEARCH BACKGROUND

As Japan is one of the countries that is most exposed to earthquakes in the world, its government has amended the laws and regulations that apply to the structural designs of buildings in response to each major earthquake occurrence that left many buildings severely damaged. Furthermore, many researchers and technologists in the country have developed and implemented a wide range of vibration control devices, seismic isolation systems, etc. and greatly improved the performance capabilities of buildings. Owing to such unrelenting efforts, the incidence of damage impacting buildings as a direct result of earthquakes has been declining.

The year 1981 was a watershed year that saw the country's laws and regulations change drastically relating to the structural designs of buildings. When the Great Hanshin Earthquake struck Japan in 1995, many buildings collapsed and caused significantly more casualties than there would have been had those buildings been able to endure the event. Indeed, it was discovered that most of those collapsed buildings were designed and constructed prior to the 1981 amendment of the Building Standards Act and the Order for Enforcement of the Building Standards Act. Hence, it became a crucial project to diagnose those old buildings to determine their seismic performance and renovate them accordingly so that they could withstand earthquakes of similar magnitude in the future.

For this purpose, the government implemented a series of measures in the same year, including the enforcement of the Act on Promotion of Seismic Retrofitting of Buildings. Under this new statute, it became a legal obligation to perform seismic diagnosis on certain structures, which were mainly buildings above a specific scale used by large numbers of people as well as storage and processing facilities where hazardous materials were handled above specific amounts.[1]

Meanwhile, the statute did not make such seismic diagnosis mandatory for factories, office buildings, and other similar structures. When it comes to seismic diagnosis performed in this context, the standard

DOI: 10.1201/9781003290261-10

procedure typically focuses on the evaluation of a building's performance using static assessment techniques based on the seismic index of structure (I_s) and ultimate horizontal resistant force (q) as stipulated in the seismic diagnosis standard, etc.[1] If it is determined that a building does not possess enough seismic performance, the building must undergo seismic retrofitting.

Since 2011, during which the Tohoku earthquake and tsunami occurred, more attention has been paid to the practice of business continuity planning (BCP) as it has become increasingly crucial for any business operator to minimize damage during a major earthquake and continue its activities afterwards. In this connection, increasing numbers of vibration control and seismic isolation devices have been implemented in existing structures in recent years for seismic reinforcement. In addition, although factories, office buildings, and other similar structures have been exempted from mandatory seismic diagnosis, more cases of voluntary seismic diagnosis and seismic reinforcement have been reported.[2, 3]

Of all the steel-framed structures that were designed prior to the 1981 statutory amendment, many factories and warehouses apparently possess lower stiffness than the steel-framed structures designed post-amendment. Therefore, the structures designed before that year are considered to exhibit larger response displacement when exposed to an earthquake, wind, or other external disturbance.

Hence, when a seismic retrofit is performed on an existing structure, it usually entails installation of braces or other reinforcement for added stiffness in addition to vibration control devices for improved seismic attenuation in order to suppress the structure's overall response displacement. However, such retrofit process requires installation of individual reinforcement materials on a number of different frames in the structure, which gives rise to a whole range of issues related to technical implementation and production activities in the form of a longer work period and increased costs resulting from an increased number of reinforcement points, and inconvenience of use due to less available openings. Furthermore, as factories typically have large machinery on the premises, it becomes necessary to secure sufficient openings for forklifts and other vehicles to pass through for normal production activities, which limits the availability of reinforcement points that can be retrofitted with ease. Moreover, if a retrofit project entails separate installation of reinforcement materials and vibration control devices, it presents various design issues in the form of a more complicated reinforcement design process requiring eccentricity ratio adjustment, etc.[4] and less energy absorption capacity resulting from the increased slender ratios of braces if the factory architecture has wider intercolumn spans and longer story heights.[5, 6]

8.2 PREVIOUS RESEARCH

8.2.1 Research on viscoelastic dampers and hybrid dampers

One of the techniques available for improving a structure's stiffness and seismic attenuation is the installation of dampers that incorporate visco-elastic materials.[7] However, with conventional viscoelastic materials, their stiffness and seismic attenuation properties tend to fluctuate with temperature. Since many old factories do not have sophisticated temperature control involving air-conditioning and building insulation materials, it is difficult for them to maintain the temperature between 10° and 30°,[8] which is the normal operating range of viscoelastic materials. Also, if the added stiffness provided to a factory by viscoelastic dampers becomes high enough relative to the factory's story stiffness, the response value – which is calculated in each analysis case reflecting the performance fluctuation of the viscoelastic materials due to temperature change – becomes more widely varied, which further complicates the design process. In addition, if viscoelastic dampers are used to retrofit a structure, the forces that the dampers generate during an earthquake of a higher magnitude than what was anticipated in the design phase might become too powerful and could cause excessive stress to the building components to which the dampers are mounted. To address this issue, however, Kasai et al.[9-11] have invented a damper design in which viscoelastic and elastoplastic materials are serially joined. This damper configuration compensates for each material's disadvantages, i.e., the viscoelastic material's likelihood of causing excessive stress to its mounted components and the elastoplastic material's inability to manifest sufficient vibration control during minor to medium-intensity earthquakes and wind-induced vibrations of low excitation amplitudes. In addition, Maseki, Narihara et al.[12] have developed this hybrid damper design that parallelly combines a damper unit of serially joined viscoelastic and elastoplastic materials with buckling-restrained braces. Elsewhere in the world, Roh, Hur et al.[13] have developed a damper design serially combining steel elements, elastoplastic elements, and lead rubber bearings (LRB) while Nasab and Kim[14] have created yet another damper configuration in which a steel slit damper and a viscoelastic element are parallelly joined. However, as the aforementioned damper designs still do not yet fully address the issue of temperature-induced performance fluctuation seen in viscoelastic materials, it might be difficult to apply them to old factories, etc. that do not have sufficient temperature control or structural stiffness for seismic retrofitting, as each of those recently introduced damper designs still utilize viscoelastic materials that are prone to significant temperature-induced performance fluctuation.

In approaching this lingering issue, especially for the seismic retrofitting of factories and the like, it is crucial to design a damper that can provide added stiffness and seismic attenuation all on its own, with little temperature-induced performance fluctuation. In terms of relative immunity to temperature-induced performance fluctuation, oil dampers are such devices that typically fit the description,[15] so it is one way of approaching the issue to combine an oil damper with a coil spring and steel elements to improve stiffness. However, if coil springs with sufficient enough stiffness are to be implemented for structural reinforcement, the device size might become significantly large, which could create a different set of issues.[16] Also, if any steel element is introduced to the damper design in combination with other components, the damper loses its stiffness-supporting capacity the moment the steel element fails, in which case the structure is no longer able to sustain its overall stiffness in response to repetitive vibrations, which poses another issue.

8.2.2 Research on seismic index of structure (I_s)

As explained in Section 8.1, when it comes to the performance evaluation of any structures that were designed and built before the 1981 amendment of the Building Standards Act and the Order for Enforcement of the Building Standards Act, it is standard practice to assess them by employing static assessment techniques based on the seismic index of structure (I_s) and ultimate horizontal resistant force (q) as specified in the seismic diagnosis standard, etc. However, in recent years, it has become increasingly popular in building reinforcement projects to utilize reinforcement materials that possess energy absorption capacity. So when taking into account this energy absorption capacity of reinforcement materials as a new factor in assessing the seismic performance of structures, it is necessary to perform time history response analysis. For the practice of seismic diagnosis to further proliferate, it is necessary to compute the seismic performance of structures, in which energy-absorbing reinforcement materials are used, in a simplified way that reflects their energy absorption capacity, and different evaluation techniques have been introduced for this objective. Kuramoto, Iiba et al.[17] have proposed one such evaluation technique that involves calculation of response and limit strength. With this particular technique, the attenuation correction factor computed by the calculation of response and limit strength in reference to the response spectrum is considered an indicator of response reduction efficacy and is evaluated as such. These researchers are proposing such method of calculating a converted I_s value by multiplying I_s by the reciprocal of the aforementioned attenuation correction factor to reflect the energy absorption capacity of reinforcement materials. Fukushima[18] applied the aforementioned technique proposed by Kuramoto et al. to compute converted I_s values while considering the effect of the added loads and energy absorption capacity of the reinforcement materials in response reduction in such a manner that would resolve the issue of redundantly considering the reinforcement effect of the reinforcement materials twice. Further, as the attenuation correction factor computed by the calculation of response and limit strength would assume steady-state response, the researcher also proposed a new technique employing a factor that considered and mitigated the effect of transient response present during earthquakes, etc. As far as quantification of the effect of transient response related to attenuation correction factor, Kasai et al.[19, 20] proposed a computation formula based on the statistics of 31 actual earthquakes previously observed, and it has been reported that the formula tracks well the actual data of those 31 seismic events. In other studies, groups led by Matsumoto, Kitajima et al.[21] and Fujii, Kitamura et al.[22] proposed different methods of calculating converted I_s values based on energy balance. In addition, another group led by Kobayashi, Inden et al.[23] proposed a modified version of the aforementioned energy-balance-based method that could calculate converted I_s values while considering the concentration of damage to certain stories in a multiple degrees-of-freedom model. Meanwhile, in another study conducted by Sato, Kasai et al.,[24] seismic response analysis data was used to estimate a seismic performance improvement factor, by which I_s would be multiplied to determine converted I_s. This method is considered highly versatile for use in wide-ranging engineering projects, as it takes into consideration the varied responses of the different stories of a reinforced building resulting from the coupled vibrations of vertically adjacent stories and can be applied to vibration control systems comprised of various reinforcement materials.

8.3 RESEARCH OBJECTIVES

In this research project, stiffness-supported oil dampers were developed with the aim of implementing them for the seismic reinforcement of factories and other similar structures that would require vibration control devices capable of stiffness improvement and seismic attenuation by themselves while presenting little temperature-induced performance fluctuation.[25] These are hybrid dampers that parallelly combine oil dampers with units each comprised of a shear-resistant spring – consisting of a low-loss viscoelastic material mostly insusceptible to temperature-induced shear failures resulting from stiffness fluctuation[26] and a steel plate – and a friction mechanism, joined serially. In addition to the typical functions of the conventional oil dampers, these dampers are capable of suppressing the deformation of the structures in which they are used, due to the shear-resistant property of the springs, even when exposed to rare seismic events as well as constant vibrations caused by normal winds. This damper configuration, where the friction mechanism and the viscoelastic spring are serially joined, also allows the friction mechanism to be set to a higher threshold so that it gets activated and creates its friction force only when exposed to an earth-quake of very rare occurrence, thereby limiting the force created by the viscoelastic spring and preventing any excessive stress from being caused to the structural components to which the viscoelastic spring is mounted. For the rest of this dissertation, this particular type of damper will be referred to simply as "spring oil damper(s)." In this connection, the aforementioned low-loss viscoelastic material's characteristics data indicating their dependence on temperature and vibration frequency are also provided in an appendix which is attached at the end. To the best of the authors' knowledge, there have been no reported cases of hybrid dampers incorporating oil dampers along with viscoelastic elements, with the purpose of the latter being to add extra stiffness to the structure instead of energy absorption, in a manner similar to how our spring oil damper functions.

Now, as for the specific objectives of this research project, they include observation of the basic characteristics of the newly developed spring oil dampers, proposal of analysis models, verification of their validity, and demonstration of the efficacy of these spring oil dampers, through the procedure described below.

(1) Conduct performance tests of the spring oil dampers by themselves and determine their basic characteristics, durability, and the aforementioned dependence (Chapter 9).

(2) Propose analysis models applicable to the spring oil dampers and compare their performance with the results of the tests from Chapter 9 to verify the validity of the analysis models (Chapter 10).

(3) Using a single degree-of-freedom model of an actual factory building, implement various reinforcement components such as steel braces, oil dampers, spring oil dampers, etc. to try out different reinforcement configurations and compare that with the results of time history response analysis to demonstrate the efficacy of the spring oil dampers (Chapter 11).

(4) Perform static assessment techniques based on the seismic index of structure (I_s) and ultimate horizontal resistant force (q) as specified in the seismic diagnosis standard, etc. to demonstrate how the response reduction efficacy of the spring oil dampers due to their seismic attenuation can be evaluated in a simplified manner (Chapter 12).

8.4 DISSERTATION COMPOSITION

Chapter 8 Preface
Section 8.1 provides the background of this research project while Section 8.2 explains about some other key studies previously conducted on viscoelastic dampers, hybrid dampers, and the seismic index of structure (I_s). Section 8.3 then sets forth the objectives of this research project.

Chapter 9 Configuration and characteristics of stiffness-supported oil dampers

This chapter explains the configuration of spring oil dampers and the results of their standalone performance tests. Section 9.2 describes the spring oil damper configuration. Section 9.3 then sets forth the basic concept and advantages of structural reinforcement using spring oil dampers. Section 9.4 provides the results of standalone performance tests performed on the stiffness-support units comprised of viscoelastic springs and friction mechanisms. Section 9.5 presents the results of standalone performance tests performed on spring oil dampers.

Chapter 10 Analysis model for stiffness-supported oil dampers

This chapter presents analysis models for spring oil dampers and compares their analysis results with the test results from Section 9.5. First, Section 10.2 describes the analysis models for spring oil dampers. Section 10.3 then compares the test results from Section 9.5 with the analysis results computed by entering the displacement waveform data obtained from the aforementioned tests into the proposed analysis models with the aim of demonstrating the validity of the analysis models.

Chapter 11 Examination involving time history response analysis based on single degree-of-freedom model

This chapter focuses on the time history response analysis that was performed on a single degree-of-freedom model mimicking an actual factory building to demonstrate the efficacy of the spring oil dampers. First, Section 11.2 provides information on the analysis model specifications and analysis conditions. Then, Section 11.3 provides analysis results and demonstrates the efficacy of the spring oil dampers.

Chapter 12 Converted I_s and q values of buildings incorporating stiffness-supported oil dampers

This chapter presents simplified methods of evaluating the response reduction efficacy of the spring oil dampers through their seismic attenuation in applying the static assessment techniques based on the seismic index of structure (I_s) and ultimate horizontal resistant force (q) as specified in the seismic diagnosis standard, etc. First, Section 12.2 provides the building model specifications. Then, Section 12.3 explains how converted I_s and q values can be calculated in a manner that reflects the response reduction efficacy of the spring oil dampers due to their seismic attenuation. Section 12.4 explains how converted I_s and q values can be calculated on the same frame model that was previously described in Section 12.2, but with the spring oil dampers installed, and compares them with the results of time history response analysis to demonstrate the validity of the converted I_s and q values that were calculated by the method proposed.

Chapter 13 Conclusion

This chapter provides a summary of the entire dissertation.

REFERENCES

1. Japan Building Disaster Prevention Association (JBDPA): *Guidelines and Commentaries on the Seismic Diagnosis and Seismic Retrofitting of Existing Steel-Framed Structures for Compliance with the Act on Promotion of Seismic Retrofitting of Buildings*, 2011 revision. (in Japanese)
2. Y Yanagawa, H Niimi, H Idota: "Seismic reinforcement design incorporating viscous dampers for large steel-framed factories", *Summaries of Scientific Lectures from the Architecture Institute of Japan (AIJ) Annual Meetings, Structure III*, pp.883–884, July 2016. (in Japanese)
3. Y Toyoda, T Nakajima, Y Suzuki, T Fujimura, Y Matsuno, S Okamoto: "Study on the effects of oil dampers in seismic retrofitting, part 1: Trial design", *Summaries of Scientific Lectures from the Architecture Institute of Japan (AIJ) Annual Meetings, Structure IV*, pp.601–602, July 2014. (in Japanese)
4. Y Yamazaki, K Kasai, H Sakata, Y Ooki: "Torsional seismic response control by viscoelastic damper for three-dimensional wood-frame structure with stiffness eccentricity", *Journal of Structural and Construction Engineering*, Architecture Institute of Japan (AIJ), Vol. 75, No. 655, pp.1691–1700, Sep. 2010. (in Japanese)
5. M Nakahara, T Shigenobu, A Watanabe: "Seismic response and restoration force characteristics of bracing structures, part 1: Equivalent elasto-plastic-slip system", *Summaries of Scientific Lectures from the Architecture Institute of Japan (AIJ) Annual Meetings, C, Structure II*, pp.1037–1038, July 1988. (in Japanese)
6. A Watanabe, T Shigenobu, M Nakahara: "Seismic response and restoration force characteristics of bracing structures, part 2: Distributed-element-type restoration force model", *Summaries of Scientific Lectures from the Architecture Institute of Japan (AIJ) Annual Meetings, C, Structure II*, pp.1039–1040, July 1988. (in Japanese)
7. S Soda, Y Takahashi: "Role of increased damping with viscoelastic dampers in seismic design of buildings", *Journal of Structural and Construction Engineering*, Architecture Institute of Japan (AIJ), No. 528, pp.67–73, Feb. 2000. (in Japanese)
8. The Japan Society of Seismic Isolation (JSSI): *Passive Seismic Control Structure Design and Construction Manual*, 2nd Edition, July 2007. (in Japanese)
9. K Kasai, M Teramoto, Y Watanabe: "Behavior of a passive control damper combining visco-elastic and elasto-plastic devices in series", *Journal of Structural and Construction Engineering*, Architecture Institute of Japan (AIJ), No. 556, pp.51–58, June 2002. (in Japanese)
10. K Kasai, Y Watanabe, N Minato: "Study on dynamic behavior of a passive control system with visco-elasto-plastic damper", *Journal of Structural and Construction Engineering*, Architecture Institute of Japan (AIJ), No. 588, pp.87–94, Feb. 2005. (in Japanese)

11. K Kasai, N Minato, K Sakurai: "Passive control design method based on tuning of equivalent stiffness of visco-elasto-plastic dampers", *Journal of Structural and Construction Engineering*, Architecture Institute of Japan (AIJ), No. 618, pp.23–31, Aug. 2007. (in Japanese)

12. R Maseki, H Narihara, Y Kimura, Y Isshiki: "Experimental study on hybrid damper composed of visco-elastic element and buckling-restrained brace element connected in parallel", *Journal of Structural Engineering*, Architecture Institute of Japan (AIJ), Vol. 57B, pp.271–278, March 2011. (in Japanese)

13. Ji-Eun Roh, Moo-Won Hur, Hyun-Hoon Choi, Sang-Hyun Lee: "Development of a multiaction hybrid damper for passive energy dissipation", *Shock and Vibration*, Vol. 2018, Jan. 2018, Article ID 5630746.

14. Mohammad Seddiq Eskandari Nasab, Jinkoo Kim: "Seismic retrofit of structures using hybrid steel slit-viscoelastic dampers", *Journal of Structural Engineering*, Vol. 146, Issue 11, Nov. 2020.

15. Y Tsuyuki, M Ogura, T Kawai, T Nakano, O Takahashi, T Okabe, S Soda, T Nakagomi: "Development of high-damping base isolation system for houses, Part 3: Oil damper design and elemental experiment", *Summaries of Scientific Lectures from the Architecture Institute of Japan (AIJ) Annual Meetings*, B-2, Structure II, pp.531–532, July 2003. (in Japanese)

16. M Takaki, S Fujii, S Soda: "Vibration isolation system for buildings using helical spring units and viscoelastic dampers", *Journal of Structural and Construction Engineering*, Architecture Institute of Japan (AIJ), No. 528, pp.83–90, Feb. 2000. (in Japanese)

17. H Kuramoto, M Iiba, A Wada: "A seismic evaluation method for existing reinforced concrete buildings retrofitted by response controlling technics", *Journal of Structural and Construction Engineering*, Architecture Institute of Japan (AIJ), No. 559, pp.189–195, Sep. 2002. (in Japanese)

18. I Fukushima: "A consideration on seismic index of structure (I_s-Index) of existing R/C buildings retrofitted with response control devices", *Summaries of Scientific Lectures from the Architecture Institute of Japan (AIJ) Annual Meetings, Structure IV*, pp.323–324, July 2012. (in Japanese)

19. K Kasai, Y Fu, A Watanabe: "Two types of passive control systems for seismic damage mitigation", *Journal of Structure Engineering*, ASCE Vol. 124, pp.501–512, May 1998.

20. K Kasai, H Ito, A Watanabe: "Peak response prediction rule for a SDOF elasto-plastic system based on equivalent linearization technique", *Journal of Structural and Construction Engineering*, Architecture Institute of Japan (AIJ), No. 571, pp.53–62, Sep. 2003. (in Japanese)

21. M Matsumoto, K Kitajima, M Nakanishi, H Adachi: "A study on seismic retrofit design of exiting R/C buildings by means of friction damper", *Proceedings of the Japan Concrete Institute*, Vol. 21, No. 1, pp.391–396, June 1999. (in Japanese)

22. K Fujii, H Kitamura, T Ochiai: "Seismic performance evaluation of existing R/C buildings with hysteresis dampers based on energy balanced response", *Journal of Structural and Construction Engineering*, Architecture Institute of Japan (AIJ), No. 613, pp.89–96, March 2007. (in Japanese)

23. M Kobayashi, T Inden, Y Isozumi, T Hasegawa, H Kitamura: "Conversion I_s index of seismic retrofitted buildings with hysteretic dampers based on the energy balance method", *Journal of Structural and Construction Engineering*, Architecture Institute of Japan (AIJ), Vol. 76, No. 663, pp.881–890, May 2011. (in Japanese)

24. T Satou, T Kasai, M Kubota, T Yamashita, S Ito, H Sakata, K Kitajima, H Tomatsuri, K Takahashi, S Shimizu, Y Yanagawa, Y Yamasaki, Y Okamoto, T Inubushi, Y Shima: "Evaluation of retrofitted building aided damping system with amplifier mechanism (Evaluation of the dynamic performance improvement magnification for the existing building retrofitted including passive control systems)", *The AIJ Journal of Technology and Design*, Vol. 19, No. 42, pp.465–470, June 2013. (in Japanese)

25. A Yokoyama, O Takahashi, Y Asano: "Development of new oil damper with spring for architectural vibration control and experimental research on structural characterization", *Summaries of Scientific Lectures from the Architecture Institute of Japan (AIJ) Annual Meetings, Structure II*, pp.145–146, July 2016. (in Japanese)

26. Taica Corporation Website (Retrieved Feb. 2, 2023 from below.) https://taica.co.jp/gel/product/shock_absorption/theta_sheet.html.

Chapter 9

Configuration and characteristics of stiffness-supported oil dampers

9.1 INTRODUCTION

This chapter explains the configuration of spring oil dampers and presents the results of their standalone performance tests. Section 9.2 describes the spring oil damper configuration. Section 9.3 then sets forth the basic concept and advantages of structural reinforcement using spring oil dampers. Section 9.4 provides the results of standalone performance tests performed on the stiffness-support units comprised of viscoelastic springs and friction mechanisms. Section 9.5 presents the results of standalone performance tests performed on spring oil dampers.

9.2 CONFIGURATION OF STIFFNESS-SUPPORTED OIL DAMPERS

Figure 9.1 illustrates the spring oil damper configuration. As indicated in the figure, the spring oil damper is a hybrid damper in which an oil damper is placed in the middle, flanked by a unit comprised of a spring made of viscoelastic material (hereinafter referred to as "viscoelastic spring(s)") and a friction mechanism on each side. This stiffness-supporting unit will be referred to simply as "spring unit(s)" for the rest of this dissertation.

These dampers are capable of suppressing the deformation of the structures in which they are used, due to the shear-resistant force of the springs, even when exposed to rare seismic events as well as constant vibrations caused by normal winds, while performing the conventional functions of oil dampers.

This damper design, which serially combines the friction mechanism and the viscoelastic spring in each spring unit, also allows the friction mechanism to be set to a higher threshold so that it gets activated and creates its friction force only when exposed to an earthquake of very rare occurrence, thereby limiting the force created by the viscoelastic spring and preventing any excessive stress from being caused to the structural components to which the viscoelastic spring is mounted. In addition, as the maximum

DOI: 10.1201/9781003290261-11

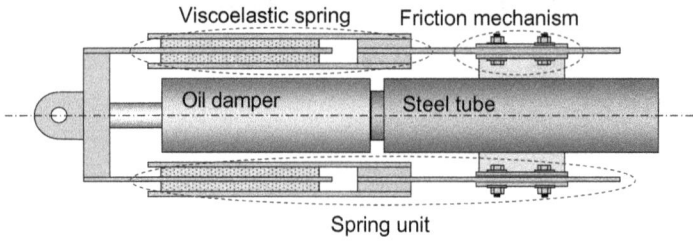

Figure 9.1 Spring oil damper configuration.

shear strain of the viscoelastic material is kept constant by the activation of the friction mechanism, this damper configuration eliminates the need to consider the maximum displacement of the viscoelastic springs during the design phase, and even if any external force greater than expected arises, the maximum shear strain of the viscoelastic material can never be surpassed. The basic specifications of the spring oil damper developed in this project are provided in Table 9.1. This particular model of oil damper has a maximum load capacity of 500 kN with the friction force set to 250 kN, resulting in the maximum force of 750 kN that the spring oil damper is able to generate. As indicated in Table 9.1, two different versions of the damper were developed in this project, one being the two-layer type in which there are a total of two layers of the viscoelastic material, and the other being the

Table 9.1 Spring oil damper specifications

Spring oil damper		Item/Type	Two-layer type	Four-layer type
		Maximum load [kN]	750	
		Maximum stroke [mm]	±80	
	Spring unit	Stiffness [kN/mm]	25	50
		Maximum displacement (spring) [mm]	10	5
		Friction force [kN]	250	250
	Oil damper	Maximum load [kN]	500	
		Maximum velocity [mm/s]	300	
		Relief load [kN]	400	
		Relief velocity [mm/s]	32	
		Primary attenuation coefficient [kN·s/mm]	12.5	
		Secondary attenuation coefficient [kN·s/mm]	0.37	
		Stiffness [kN/mm]	140	

four-layer type having a total of four layers of the same material. Because the maximum displacement of the viscoelastic spring is 5 mm for the two-layer type and 10 mm for the four-layer type, with the thickness of each viscoelastic material layer being 10 mm, the maximum shear strain is 100% for the two-layer type and 50% for the four-layer type.

9.3 BASIC CONCEPT AND ADVANTAGES OF REINFORCEMENT WITH STIFFNESS-SUPPORTED OIL DAMPERS

As explained above, when it comes to performing seismic retrofits involving oil dampers on factories and other structures designed before the 1981 statutory amendment, it is common practice to use steel-frame braces for improving stiffness. As the spring oil dampers that have been developed in this project are able to improve stiffness and seismic attenuation all on their own, they can reduce the number of reinforcement points compared to reinforcement involving conventional braces and oil dampers. Hence, these oil dampers allow many openings to be left available for use after reinforcement. In addition, as they reduce the number of reinforcement points, they can not only save the cost of reinforcement components but also the cost of reinforcement related to columns, beams, footing beams, etc. Figure 9.2 through Figure 9.4 provide illustrations of how seismic reinforcement can be achieved using reinforcement components such as braces, oil dampers, and spring oil dampers.

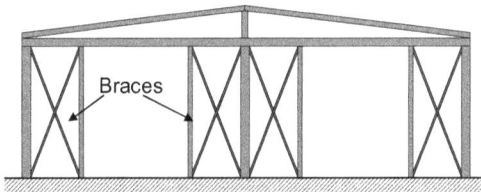

Figure 9.2 Basic concept of seismic reinforcement with braces.

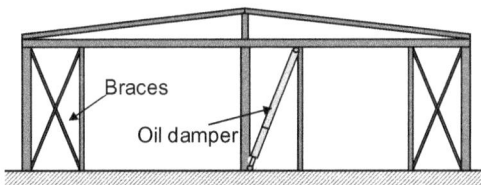

Figure 9.3 Example of seismic reinforcement with braces and oil damper.

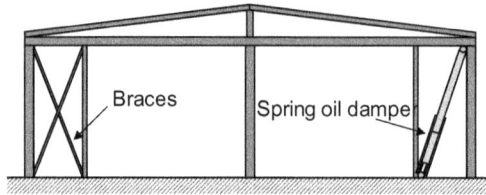

Figure 9.4 Example of seismic reinforcement with spring oil damper.

9.4 DYNAMIC TEST OF STIFFNESS SUPPORT DEVICES

This section describes the tests that were conducted on the spring units – which had been developed for stiffness support – to determine the basic characteristics, multi-factor dependence, and durability of the viscoelastic springs and friction mechanism comprising the unit, and their results.

9.4.1 Test object specifications

For this dynamic test of the spring units by themselves, the test objects were the two types of spring oil dampers with the stiffness of their viscoelastic springs set to 25 kN/mm and 50 kN/mm as indicated in Table 9.1. Before this test, the oil dampers were drained of hydraulic fluid so that no force would be generated by them in order to precisely determine the performance characteristics of each of the viscoelastic springs and friction mechanism.

9.4.2 Test outline

As indicated in Figure 9.5, each of the spring oil dampers as test objects was mounted to the vibration test apparatus to conduct the vibration test involving forced displacement. As for the vibration test apparatus specifications, it had a maximum vibration force of 600 kN, operation stroke of ±125 mm, vibration frequency range of 0.1 Hz to 33 Hz, and maximum velocity of

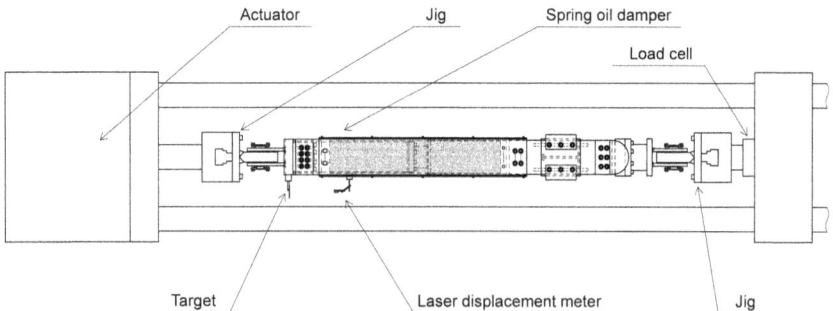

Figure 9.5 Test configuration.

720 mm/s. Test jigs were used to set the spring oil dampers to the test apparatus, along with a laser displacement meter to measure the displacement caused to the spring oil dampers. The load cell that was built into the test apparatus was used to measure the force generated by the spring oil dampers. Figures 9.6 and 9.7 provide photographic images of the test set-up.

9.4.3 Basic characteristics test of viscoelastic springs

In this segment of the test, five cycles of sine wave vibration were applied, including a cycle in the incremental phase and another in the decremental phase as indicated in Figure 9.8, to determine the basic characteristics (stiffness, material strength, and adhesive force) of the viscoelastic springs as well as their dependence on vibration frequency. The test conditions adopted are indicated in Tables 9.2 and 9.3. For both two-layer and four-layer types, the range of sine wave cycles of 0.25 to 2 sec. was applied, while four different sets of test conditions were implemented for excitation amplitudes at and below maximum displacement and another set of test conditions for excitation amplitudes above maximum displacement, again for both types. For this test procedure, to determine the performance of the viscoelastic springs by themselves, the high-strength bolts fastening each friction mechanism serially connected to the viscoelastic spring were tightened with sufficient torque so that the friction mechanism would not slide at all during the test. For the evaluation of test results, data of the

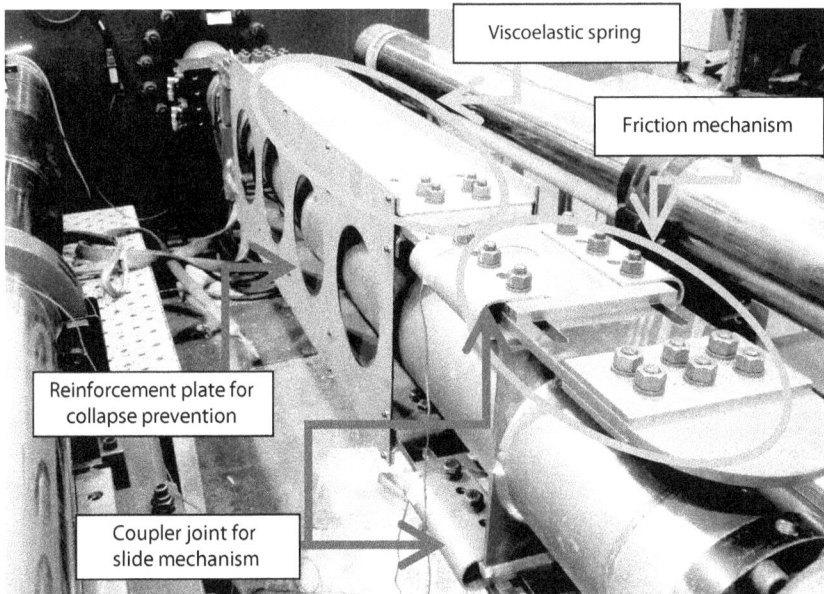

Figure 9.6 Spring oil damper tested (two-layer type).

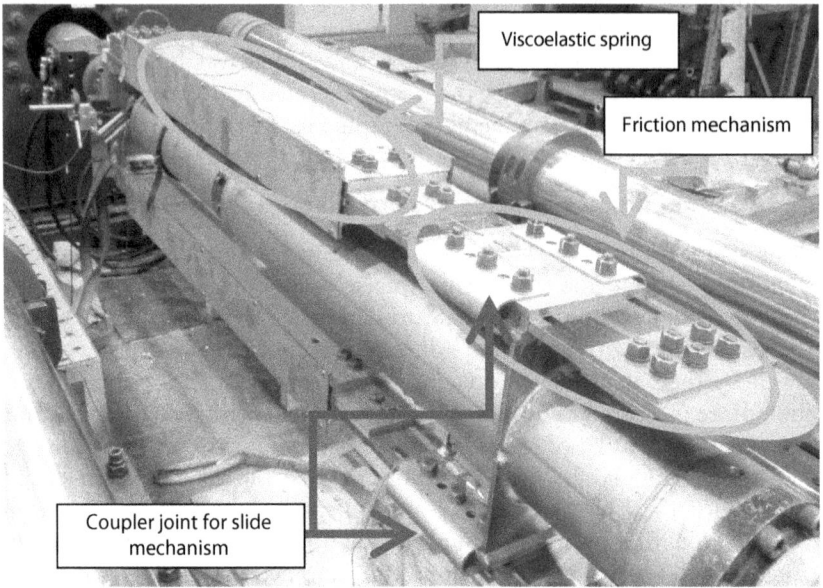

Figure 9.7 Spring oil damper tested (four-layer type).

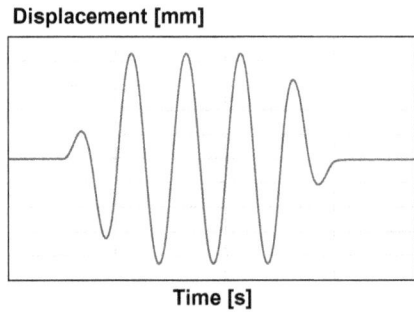

Figure 9.8 Input waveform.

Table 9.2 Conditions of the basic characteristics test of viscoelastic springs (two-layer type)

		Shear strain (displacement)				
Test object	*Cycle [s]*	*25% (2.5 mm)*	*50% (5 mm)*	*75% (7.5 mm)*	*100% (10 mm)*	*120% (12 mm)*
Two-layer type	0.25	O	O	O	O	O
	0.5	O	O	O	O	O
	1	O	O	O	O	O
	2	O	O	O	O	O

Table 9.3 Conditions of the basic characteristics test of viscoelastic springs (four-layer type)

Test object	Cycle [s]	Shear strain (displacement)				
		12.5% (1.25 mm)	25% (2.5 mm)	37.5% (3.75 mm)	50% (5 mm)	70% (7 mm)
Four-layer type	0.25	O	O	O	O	O
	0.5	O	O	O	O	O
	1	O	O	O	O	O
	2	O	O	O	O	O

third sine wave cycle was sampled, while no smoothing (moving average calculation) was performed on the measured data. In terms of the sampling frequency of measured data, it was adjusted so that the number of data per waveform would be at least 400 for each vibration cycle. All measured data were processed through a 40-Hz low-pass filter. The results of this test to determine the basic characteristics of the viscoelastic springs are shown in Table 9.4 (two-layer type) and Table 9.5 (four-layer type) while a plot indicating the relationship between maximum displacement and the force measured at maximum displacement for each vibration frequency is presented in Figure 9.9. In addition, graphs indicating the relationship between the force and displacement of the viscoelastic springs under each set of the test conditions are presented in Figure 9.10 (two-layer type) and Figure 9.11 (four-layer type).

Based on analysis of the test results shown in Tables 9.4 and 9.5 and the plotted data in Figure 9.9, it was determined that the average stiffness of the two-layer and four-layer types while exposed to sine wave vibrations of the various cycles was –12.5% (two-layer-type) and –14.1% (four-layer type) of the design-specified values (two-layer type: 25 kN/mm at 100% shear strain; four-layer type: 50 kN/mm at 50% shear strain). As the viscoelastic material had displacement dependence, the stiffness apparently declined as displacement rose. Also, from the test results shown in Tables 9.4 and 9.5, the plotted data in Figure 9.9, and the graphs indicating the force-displacement relation in Figures 9.10 and 9.11, it was discerned that both stiffness and attenuation coefficient had dependence on vibration frequency. However, since the stiffness fluctuation remained roughly in the ±10% range within the vibration frequencies applied in this test, the stiffness fluctuation induced by vibration frequency dependence appeared to be minor relative to ISD111 and other viscoelastic materials[1] that are commonly used to manufacture viscoelastic dampers. In addition, while this test protocol caused a maximum strain equivalent to 120% (two-layer type) of the design-specified value related to the viscoelastic springs, no apparent damage was caused to the viscoelastic material nor did it detach from the

Table 9.4 Results of the basic characteristics test of viscoelastic springs (two-layer type)

		Test results		Design-specified values		
Cycle	Excitation amplitude	Force	Stiffness	Force	Stiffness	Deviation
[s]	[mm]	[kN]	[kN/mm]	[kN]	[kN/mm]	[%]
0.25	2.53	66.36	26.23	63.25	25	4.9
	4.92	120.54	24.50	123.00		−2.0
	7.40	172.28	23.28	185.00		−6.9
	9.94	200.84	20.21	248.50		−19.2
	11.72	237.58	20.27	293.00		−18.9
0.5	2.52	67.38	26.74	63.00		7.0
	4.93	123.62	25.08	123.25		0.3
	7.40	176.72	23.88	185.00		−4.5
	10.00	218.24	21.82	250.00		−12.7
	11.89	265.66	22.34	297.25		−10.6
1	2.51	67.82	27.02	62.75		8.1
	4.94	123.40	24.98	123.50		−0.1
	7.43	176.56	23.76	185.75		−4.9
	10.02	228.06	22.76	250.50		−9.0
	11.96	270.98	22.66	299.00		−9.4
2	2.54	67.94	26.75	63.50		7.0
	5.00	123.92	24.78	125.00		−0.9
	7.49	177.40	23.68	187.25		−5.3
	10.07	229.24	22.76	251.75		−8.9
	12.07	270.70	22.43	301.75		−10.3

steel plate when it was checked after the test. Based on the above, it was determined that the strength of the viscoelastic material and the adhesive force by which the viscoelastic material stuck to the steel plate were both sufficient.

9.4.4 Durability test of viscoelastic springs

In this segment of the test, which was conducted after the test of viscoelastic springs to determine basic characteristics, 100 cycles of sine wave vibration, including a cycle in the incremental phase and another in the decremental phase, were applied in a manner similar to the one shown in Figure 9.8 for the test of viscoelastic springs to determine basic characteristics, in order to examine the durability (material strength and adhesive force) of the viscoelastic springs that would be required in major earthquakes. In addition, following the aforementioned durability test, a vibration test applying

Table 9.5 Results of the basic characteristics test of viscoelastic springs (four-layer type)

		Test results		Design-specified values		
Cycle	Excitation amplitude	Force	Stiffness	Force	Stiffness	Deviation
[s]	[mm]	[kN]	[kN/mm]	[kN]	[kN/mm]	[%]
0.25	1.28	60.90	47.58	64.00	50	−4.8
	2.49	112.46	45.16	124.50		−9.7
	3.74	164.04	43.86	187.00		−12.3
	4.98	211.92	42.55	249.00		−14.9
	6.92	277.92	40.16	346.00		−19.7
0.5	1.29	62.52	48.47	64.50		−3.1
	2.45	114.54	46.75	122.50		−6.5
	3.75	167.12	44.57	187.50		−10.9
	5.00	215.16	43.03	250.00		−13.9
	7.00	288.62	41.23	350.00		−17.5
1	1.30	63.86	49.12	65.00		−1.8
	2.39	112.06	46.89	119.50		−6.2
	3.69	166.24	45.05	184.50		−9.9
	4.97	214.74	43.21	248.50		−13.6
	6.99	287.90	41.19	349.50		−17.6
2	1.39	67.44	48.52	69.50		−3.0
	2.34	109.38	46.74	117.00		−6.5
	3.67	164.66	44.87	183.50		−10.3
	4.96	213.24	42.99	248.00		−14.0
	7.00	287.12	41.02	350.00		−18.0

sine waves of 1-second cycles was conducted under the same conditions as those of the basic characteristics test to observe how the performance would change before and after the durability test. Table 9.6 describes the conditions under which the durability test was conducted while Table 9.7 sets forth the conditions under which the basic characteristics test was carried out after the durability test. As with the basic characteristics test, this durability test of viscoelastic springs was conducted with the friction mechanism fully fastened so that no sliding would occur.

Figure 9.12 provides the results of the durability test conducted on the viscoelastic springs, in the form of superimposition charts indicating the relationship between force and displacement that manifested during the test. In addition, Tables 9.8 and 9.9 summarize the results of the basic characteristics tests conducted before and after the durability test while Figure 9.13 presents a superimposition chart indicating the relationship between force and displacement.

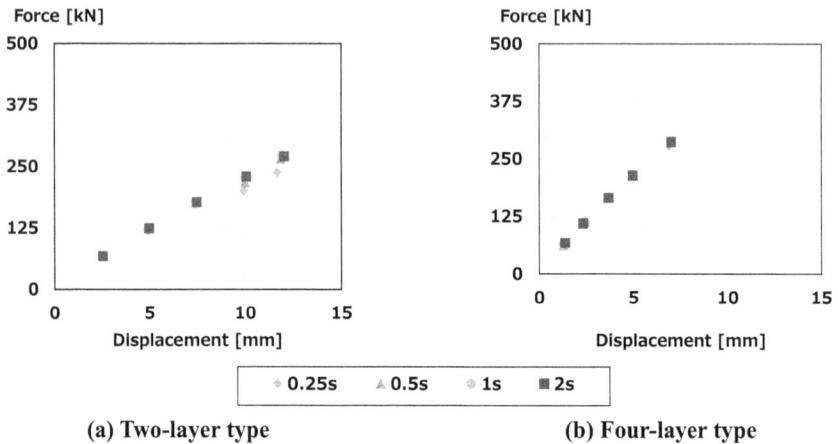

(a) Two-layer type   (b) Four-layer type

Figure 9.9 Plot indicating relation between maximum displacement and the force measured at maximum displacement: Test of viscoelastic springs to determine basic characteristics.

Observation of the force-displacement relationship as indicated in Figure 9.12 revealed that the viscoelastic springs exhibited steady performance as represented by the lines in the graphs during the durability test. From these results, it was determined that the viscoelastic material did not sustain any damage while the adhesive force that held together the viscoelastic material and the steel plate did not decrease throughout the durability test. It was also discerned that neither stiffness nor attenuation coefficient fluctuated much under the effect of vibration. Furthermore, based on the test results presented in Tables 9.8 and 9.9 and the force-displacement relationship indicated in Figure 9.13, it was determined that stiffness had hardly changed before and after the durability test. It was also confirmed after the test that no damage had been sustained by the viscoelastic material and that the viscoelastic material had not gotten detached from the steel plate. Hence, it became clear that the material strength of the viscoelastic material and the adhesive force holding together the viscoelastic material and the steel plate were both sufficiently high to withstand a major earthquake event.

9.4.5 10,000-cycle vibration test of viscoelastic springs

In this segment of the test, 10,000 cycles of sine wave vibration – including a cycle in the incremental phase and another in the decremental phase as indicated in Figure 9.8 for the basic characteristics test – were applied to the viscoelastic springs after the basic characteristics test and the durability test to determine the durability (material strength and adhesive force) of

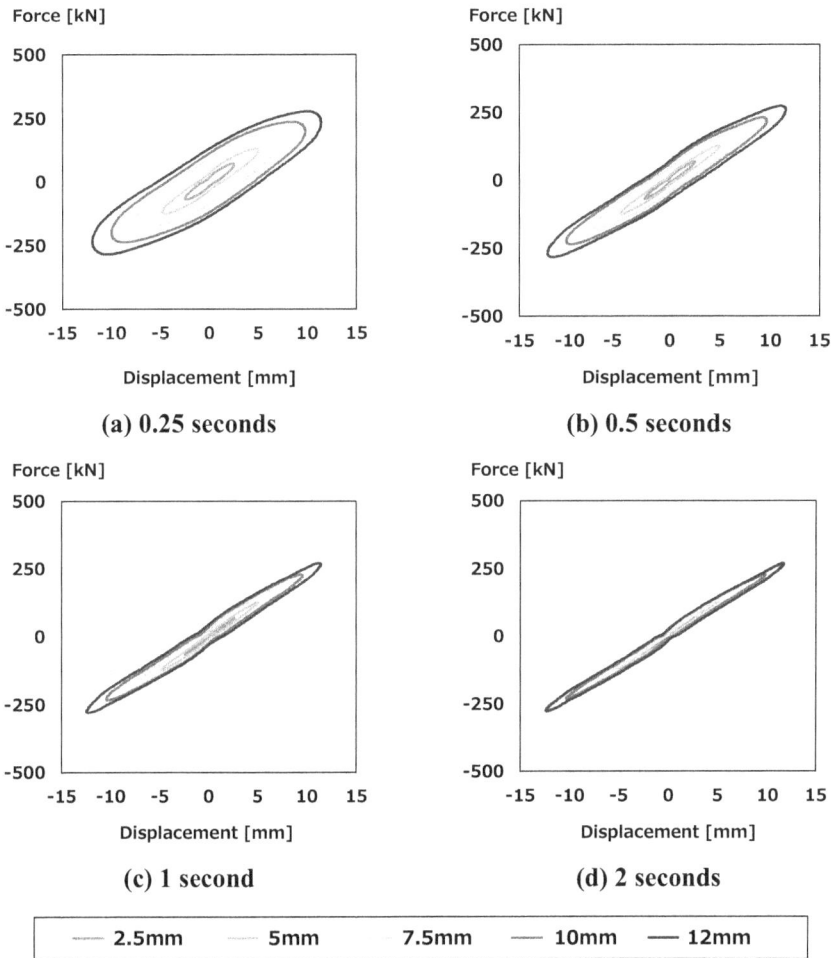

Force [kN]

(a) 0.25 seconds

Force [kN]

(b) 0.5 seconds

Force [kN]

(c) 1 second

Force [kN]

(d) 2 seconds

Displacement [mm]

—— 2.5mm   —— 5mm   7.5mm   —— 10mm   —— 12mm

Figure 9.10 Chart indicating relation between force and displacement: Test of viscoelastic springs to determine basic characteristics (two-layer type).

the viscoelastic springs under the effects of constant vibrations mimicking normal winds as well as minor to medium-intensity earthquakes. In addition, following this 10,000-cycle vibration test, a sine wave vibration test with a cycle of 1 second was conducted applying the same conditions as those applied to the test that was performed after the durability test, in order to observe how the performance would change before and after the 10,000-cycle vibration test. Tables 9.10 and 9.11 provide the conditions of these tests. As with the basic characteristics tests previously performed, this durability test of the viscoelastic springs was also conducted with the friction mechanism fully fastened so that no sliding would occur.

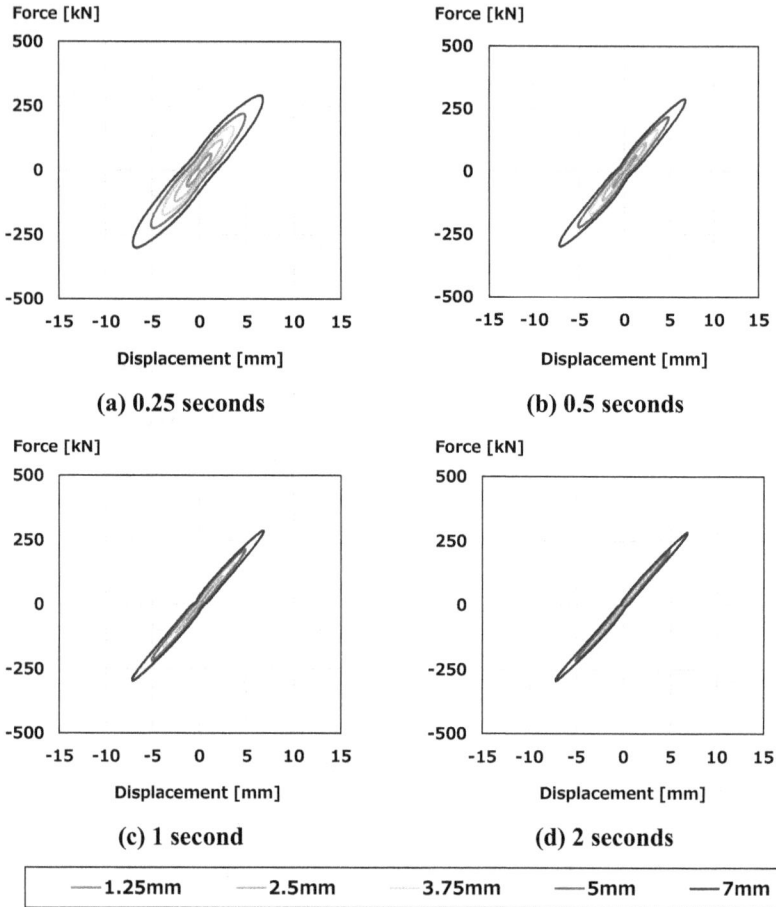

Figure 9.11 Chart indicating relation between force and displacement: Test of viscoelastic springs to determine basic characteristics (four-layer type).

Table 9.6 Conditions of the durability test of viscoelastic springs

Test object	Cycle [s]	Shear strain (displacement)	No. of cycles
Two-layer type	1	100% (10 mm)	100
Four-layer type		50% (5 mm)	

Table 9.7 Conditions of the basic characteristics test (before and after the durability test)

Test object	Cycle [s]	Shear strain (displacement)		
Two-layer type	1	50% (5 mm)	100% (10 mm)	120% (12 mm)
Four-layer type		25% (2.5 mm)	50% (5 mm)	70% (7 mm)

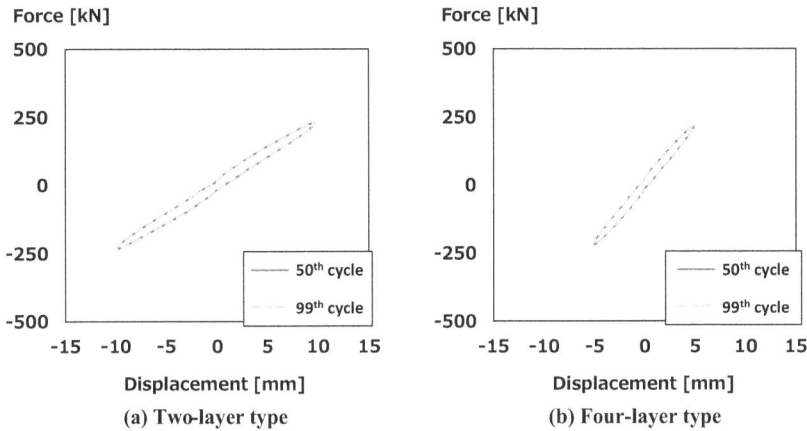

Figure 9.12 Chart indicating relation between force and displacement: Durability test of viscoelastic springs.

Table 9.8 Results of the basic characteristics test (before and after the durability test) (two-layer type)

		Test results			Design-specified value		
	Cycle	Excitation amplitude	Force	Stiffness	Force	Stiffness	Deviation
	[s]	[mm]	[kN]	[kN/mm]	[kN]	[kN/mm]	[%]
Before durability test	1	4.94	123.40	24.98	123.50	25	-0.1
		10.02	228.06	22.76	250.50		-9.0
		11.96	270.98	22.66	299.00		-9.4
After durability test		4.89	124.48	25.46	122.25		1.8
		9.94	224.18	22.55	248.50		-9.8
		12.03	273.72	22.75	300.75		-9.0

Tables 9.12 and 9.13 indicate the results of the basic characteristics tests conducted before and after the durability test while Figure 9.14 presents a superimposition chart indicating the relationship between force and displacement.

Examination of the test results provided in Tables 9.12 and 9.13 and the graphs indicating the force-displacement relation in Figure 9.14 revealed that the viscoelastic exhibited steady performance as represented by the lines in the graphs before and after the 10,000-cycle vibration test as their stiffness remained virtually constant throughout. It was also confirmed after the test that no damage had been sustained by the viscoelastic material and that the viscoelastic material had not gotten detached from the steel plate. Based on the above, it was determined the viscoelastic material possessed sufficient strength and the adhesive force that held the viscoelastic

Table 9.9 Results of the basic characteristics test (before and after the durability test) (four-layer type)

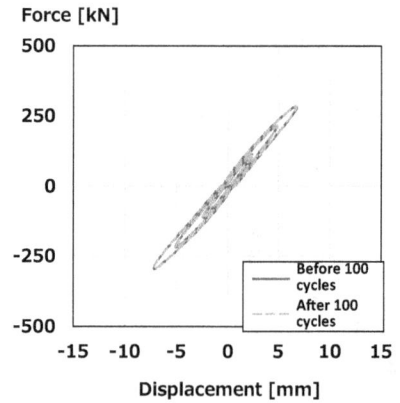

			Test results		Design-specified values		
	Cycle	Excitation amplitude	Force	Stiffness	Force	Stiffness	Deviation
	[s]	[mm]	[kN]	[kN/mm]	[kN]	[kN/mm]	[%]
Before durability test	I	2.39	112.06	46.89	119.50	50	−6.2
		4.97	214.74	43.21	248.50		−13.6
		6.99	287.90	41.19	349.50		−17.6
After durability test		2.46	115.50	46.95	123.00		−6.1
		4.99	217.02	43.49	249.50		−13.0
		6.99	288.56	41.28	349.50		−17.4

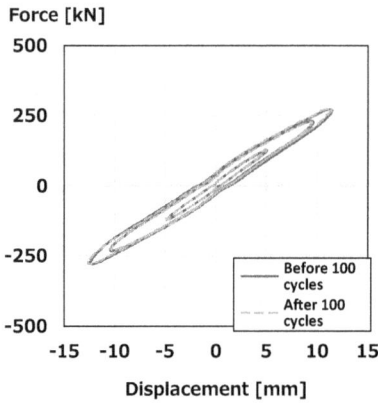

(a) Two-layer type   (b) Four-layer type

Figure 9.13 Chart indicating relation between force and displacement: Basic characteristics test (before and after the durability test).

Table 9.10 Conditions of the 10,000-cycle vibration test of viscoelastic springs

Test objects	Cycle [s]	Input displacement [mm]	No. of cycles
Two-layer type	0.5	2.5	10,000
Four-layer type			

Table 9.11 Conditions of the basic characteristics test (before and after the 10,000-cycle vibration test)

Test objects	Cycle [s]	Shear strain (displacement)		
Two-layer type	1	50% (5 mm)	100% (10 mm)	120% (12 mm)
Four-layer type		25% (2.5 mm)	50% (5 mm)	70% (7 mm)

Table 9.12 Results of the basic characteristics test (before and after the 10,000-cycle vibration test) (two-layer type)

	Cycle	Test results			Design-specified values		
		Excitation amplitude	Force	Stiffness	Force	Stiffness	Deviation
	[s]	[mm]	[kN]	[kN/mm]	[kN]	[kN/mm]	[%]
Before durability test	1	4.89	124.48	25.46	122.25	25	1.8
		9.94	224.18	22.55	248.50		−9.8
		12.03	273.72	22.75	300.75		−9.0
After durability test		4.95	126.26	25.51	123.75		2.0
		10.11	228.94	22.64	252.75		−9.4
		11.97	273.30	22.83	299.25		−8.7

Table 9.13 Results of the basic characteristics test (before and after the 10,000-cycle vibration test) (four-layer type)

	Cycle	Test results			Design-specified values		
		Excitation amplitude	Force	Stiffness	Force	Stiffness	Deviation
	[s]	[mm]	[kN]	[kN/mm]	[kN]	[kN/mm]	[%]
Before durability test	1	2.46	115.50	46.95	123.00	25	−6.1
		4.99	217.02	43.49	249.50		−13.0
		6.99	288.56	41.28	349.50		−17.4
After durability test		2.60	124.44	47.86	130.00		−4.3
		5.09	220.74	43.37	254.50		−13.3
		7.00	289.92	41.42	350.00		−17.2

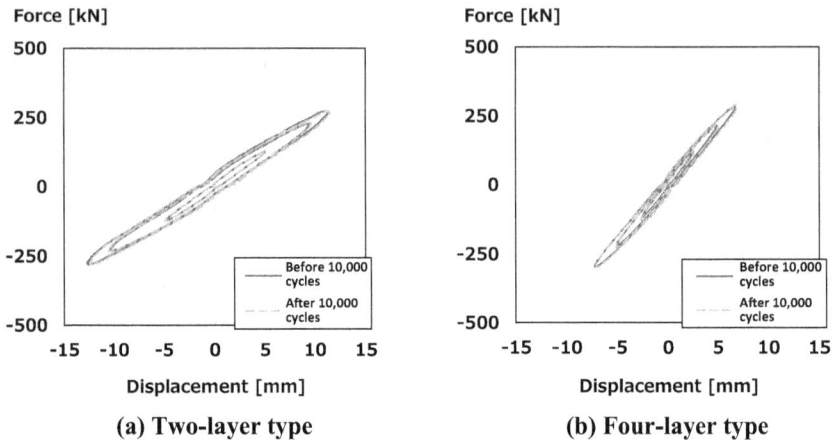

(a) Two-layer type   (b) Four-layer type

Figure 9.14 Chart indicating relation between force and displacement: Basic characteristic test (before and after the 10,000-cycle vibration test).

material and the steel plate together was sufficient to withstand the effects of constant vibrations equivalent to normal winds and minor to medium-intensity earthquakes.

9.4.6 Operation characteristics test of friction mechanism

In this segment of the test, three cycles of sine wave vibration – including a cycle in the incremental phase and another in the decremental phase as indicated in Figure 9.8 for the basic characteristics test of the viscoelastic springs – were applied to the friction mechanism to observe its operation characteristics, friction force, and its multi-factor dependence (on velocity and vibration frequency). Table 9.14 describes the test conditions.

Figure 9.15 provides the results of this operation characteristics test conducted on the friction mechanism, in the form of superimposition charts indicating the relationship between force and displacement that manifested

Table 9.14 Conditions of the operation characteristics test of friction mechanism

(a) Two-layer type				(b) Four-layer type			
	Excitation amplitude [mm]				Excitation amplitude [mm]		
Cycle [s]	20	40	60	Cycle [s]	10	30	60
0.5	○	—	—	0.5	○	○	—
1	○	○	○	1	○	○	○
2	○	○	○	2	○	○	○

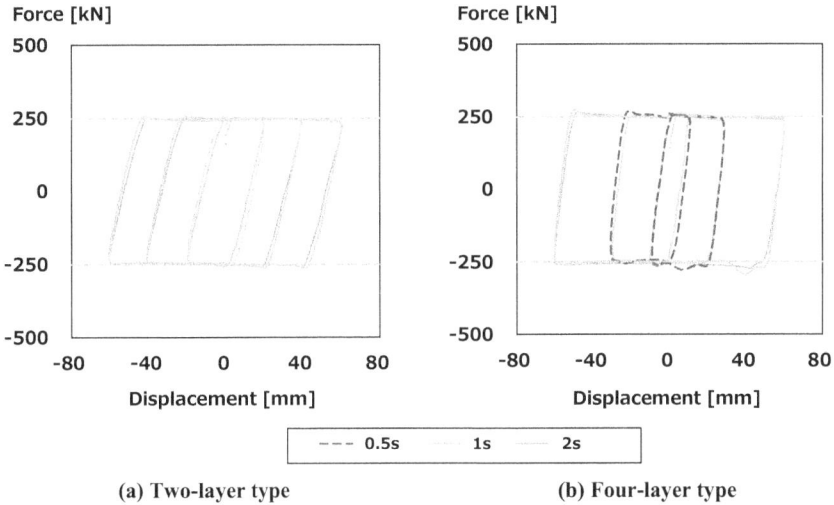

Figure 9.15 Chart indicating relation between force and displacement: Operation characteristics test of friction mechanism.

Table 9.15 Conditions of the durability test of friction mechanism

Cycle [s]	Input displacement [mm]	No. of cycles
1	30	100

under each set of the test conditions. After examination of the data presented in Figure 9.15, it was determined that the friction mechanism had operated stably under all the test conditions, consistently exhibiting a friction force of 250 kN as specified in its design, without being affected by the changes in velocity or vibration frequency.

9.4.7 Durability test of friction mechanism

In this segment of the test, 100 cycles of sine wave vibration – including a cycle in the incremental phase and another in the decremental phase as with the operation characteristics test – were applied to the friction mechanism to determine its durability. Table 9.15 provides the test conditions.

Figure 9.16 provides the results of this durability test conducted on the friction mechanism, in the form of superimposition charts indicating the relationship between force and displacement. After examination of the data presented in Figure 9.16, it was determined that the friction mechanism had operated stably as the number of cycles increased, consistently exhibiting a friction force of 250 kN as specified in its design. Based on the above, it was determined that the durability of the friction mechanism was sufficient to endure major earthquake events.

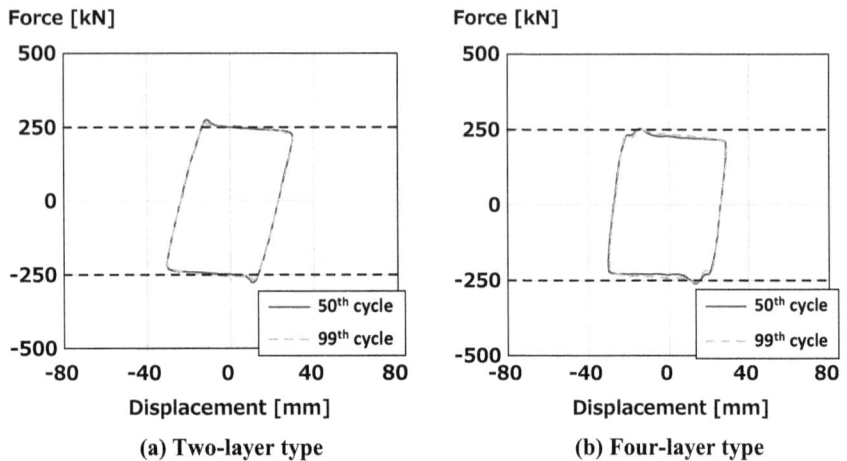

Figure 9.16 Chart indicating relation between force and displacement: Durability test of friction mechanism.

9.4.8 Random wave vibration test of spring units

In this segment of the test, random wave vibration was applied mimicking seismic waves to ensure that the viscoelastic springs and the friction mechanism would operate smoothly. Figure 9.17 indicates the displacement waveform applied in this test, which was created by conversion of the acceleration data from the JMA KOBE NS 1995 seismic event with such amplitude adjustment that would result in a maximum displacement of 60 mm. Figure 9.18 provides the results of this random wave response test, in the form of charts indicating the relationship between force and

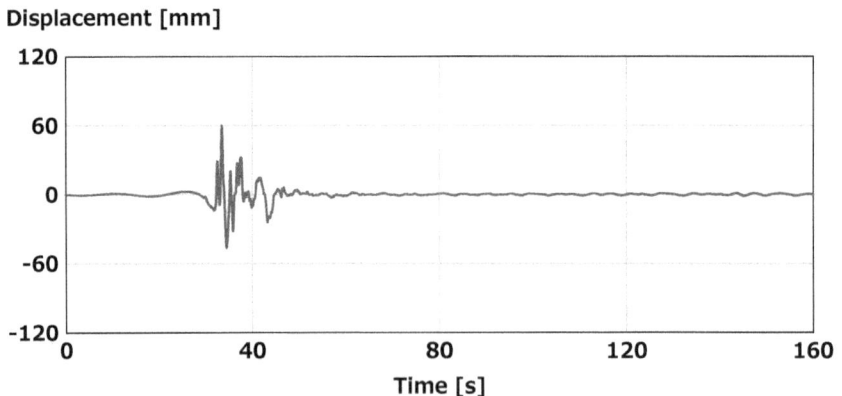

Figure 9.17 Chart indicating relation between displacement and time: JMA KOBE NS 1995, maximum displacement 60 mm.

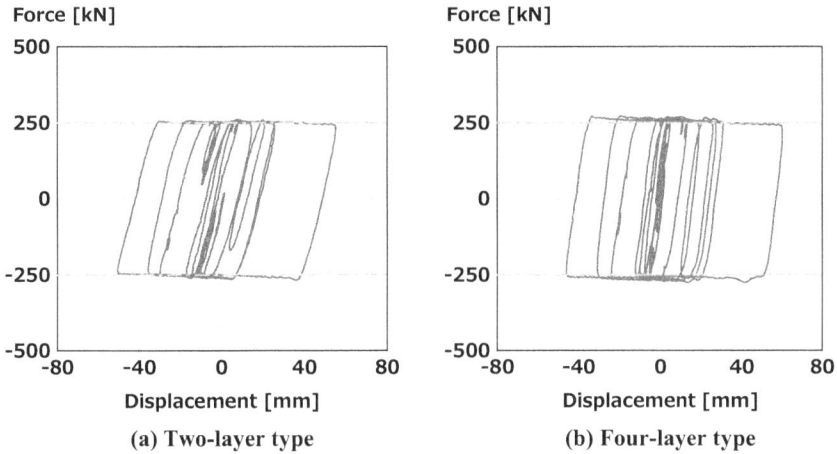

Figure 9.18 Chart indicating relation between force and displacement: Random wave vibration test of spring units.

displacement. As this test revealed that both the viscoelastic springs and the friction mechanism exhibited steady performance as represented by the lines in the graphs in response to the random wave input, it was determined that the friction mechanism had operated stably throughout, exhibiting a friction force of 250 kN consistently with its design-specified value.

9.5 PERFORMANCE TEST OF STIFFNESS-SUPPORTED OIL DAMPERS

This section describes the tests that were conducted on the stiffness-supported oil dampers to determine their basic characteristics along with their results.

9.5.1 Test object specifications

For these tests, the test object was one of the two types of spring oil dampers as indicated in Table 9.1 above comprised of the less stiff spring unit (two-layer type) and the oil damper with its maximum load capacity of 250 kN. As the maximum force that could be generated by the test apparatus was 500 kN, the maximum load of the 500 kN-limit oil damper was adjusted to 250 kN while the friction force of the friction mechanism when it got activated was set to 125 kN, resulting in a total maximum force of 375 kN that would be generated by the entire spring oil damper. Table 9.16 provides the test object specifications. Before the main test was conducted, a preliminary test was performed on the oil damper alone to

Table 9.16 Spring oil damper specifications

Spring oil damper		Maximum load [kN]	375
		Maximum stroke [mm]	±80
	Oil damper	Maximum load [kN]	250
		Maximum velocity [mm/s]	150
		Relief load [kN]	200
		Relief velocity [mm/s]	32
		Primary attenuation coefficient [kNs/mm]	6.25
		Secondary attenuation coefficient [kNs/mm]	0.42
		Stiffness [kN/mm]	140
	Spring unit	Stiffness [kN/mm]	25
		Maximum displacement (spring) [mm]	5
		Friction force [kN]	125

determine its characteristics. For this test, sine waves were applied with cycles of 4 sec. (primary attenuation coefficient phase) and 1 sec. (post-relief secondary attenuation coefficient phase). Figure 9.19 provides a plot indicating the relationship between maximum force and velocity in response to vibration input. For this purpose, velocity was calculated using the formula (9.1)[1] based on the maximum excitation amplitude applied during the test. Examination of the force-velocity relationship indicated in Figure 9.19 revealed that the oil damper had exhibited characteristics approximating the design-specified values as set forth in Table 9.16. In addition, another preliminary test was conducted to determine the characteristics of the viscoelastic spring, which was a static test entailing draining oil from the

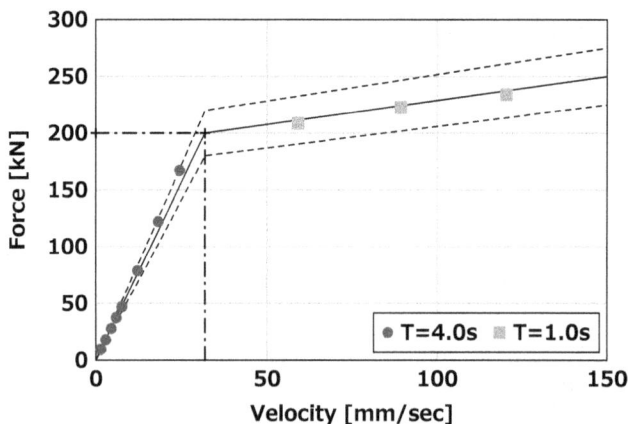

Figure 9.19 Oil damper performance test results. *The black solid line indicates design specifications while the black dashed lines indicate the range of ±10% deviation from the design specifications.

oil damper, fully tightening the high-strength bolts fastening the friction mechanism, and causing forced displacement to the spring oil damper in order to determine the force-displacement curve related to the viscoelastic spring. Figure 9.20 provides the results of the tests involving application of force in both tensile and compressive directions as well as the force-displacement curve, which was plotted by averaging the results of the two aforementioned tests involving tensile and compressive force. Examination of Figure 9.20 revealed that the viscoelastic spring had a dependence on displacement. It was also determined that the viscoelastic spring performed more or less to the design specifications, as its stiffness was set to reach 25 kN/mm when its deformation became 10 mm.

$$\dot{u}_{d,max} = u_{d,max} \cdot \dot{E} = 2 \cdot \dot{A} \cdot u_{d,max} \cdot f \qquad (9.1)$$

where $\dot{u}_{d,max}$ = maximum velocity [mm/s], $u_{d,max}$ = maximum excitation amplitude [mm], ω = circular vibration frequency [rad/s], and f = vibration frequency [Hz].

9.5.2 Test outline

As indicated in Figures 9.21 and 9.22, each of the spring oil dampers was mounted to the vibration test apparatus to conduct the vibration test involving forced displacement. As for the vibration test apparatus specifications, it had a maximum vibration force of 600 kN, operation stroke of ±125 mm, vibration frequency range of 0.1 Hz to 33 Hz, and maximum velocity of 72 cm/s. Test jigs were used to set the spring oil dampers to the test apparatus, and the measuring instrument was installed as indicated in Figures 9.23 and 9.24, along with the laser displacement meter to measure the displacement caused to the spring oil dampers. The load cell that was built into the test apparatus was used to measure the force generated by the

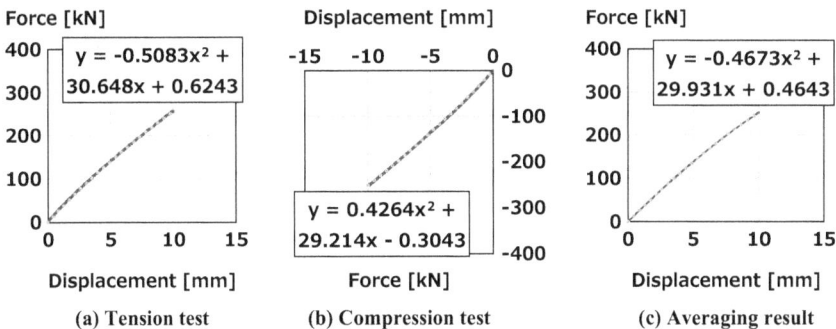

Figure 9.20 Chart indicating relation between force and displacement: Viscoelastic spring under static load.

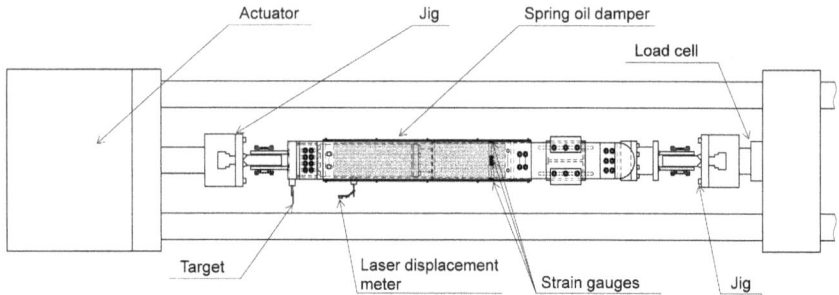

Figure 9.21 Test configuration.

Figure 9.22 Test set-up.

entire spring oil dampers. Figures 9.23 and 9.24 indicate how the strain gauges were attached to the steel tube connected to the oil damper so that the force generated by the oil damper could be measured by the reading on the strain gauges. The force generated by the spring unit was calculated by subtracting the force being applied to the oil damper from the force affecting the entire spring oil damper.

To allow for the calculation of the force bearing upon the oil damper based on the strain gauge reading, a preliminary test was conducted, which involved loosening the high-strength bolts fastening the friction mechanism, so that the force would be generated only from the oil damper, and applying sine wave vibration with a cycle of 1 second and excitation amplitude of 2.5 mm, to determine a conversion factor by which to calculate the load cell output based on the strain gauge reading. For this test procedure, the sampling frequency of measured data was 500 Hz and the data was processed through a 100 Hz low-pass filter. Then, as indicated in Figure 9.24, the bridge circuit was created between the strain gauges facing each other so that any effect of flexural strain could be removed during measurement. Each of such strain gauge combinations will be referred to simply as "Strain 1" and "Strain 2" for the rest of this dissertation. Figure 9.25 provides

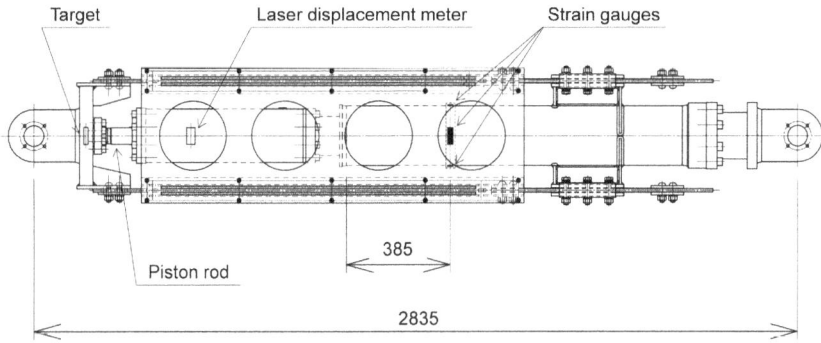

Figure 9.23 Measuring instrument set-up.

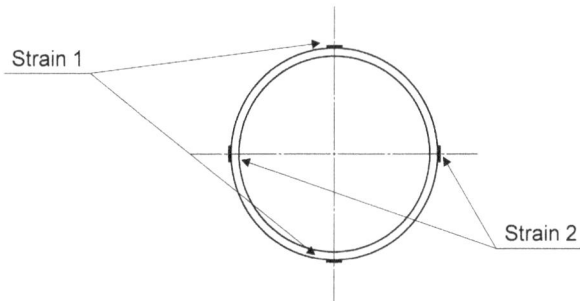

Figure 9.24 Strain gauge set-up.

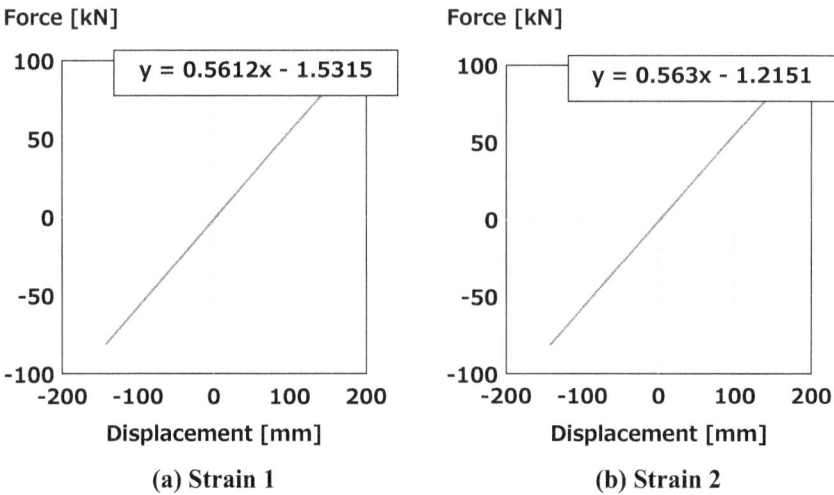

(a) Strain 1   (b) Strain 2

Figure 9.25 Chart indicating relation between force and strain.

graphs indicating the relationship between force and strain bearing upon the load cell, from which it is apparent that the force output from the load cell and the strain output from the strain gauges are of a linear relation, so it can be assumed that the output from the strain gauges was basically free of the effect of any bending strain caused by their own weight that would have caused a non-axial force had it not been removed properly. In Figure 9.25, the slope of the linear function was applied as the conversion factor by which to multiply the measured data of Strains 1 and 2, and the average of all force data measured was applied as the force generated by the oil dampers to draw the graphs. The constant terms in the formulas shown in Figure 9.25 were ignored because they would only relate to the initial values indicated on the load cell and the strain gauges.

9.5.3 Sine wave vibration test

In this segment of the test, five cycles of sine wave vibration – including a cycle in the incremental phase and another in the decremental phase like those indicated in Figure 9.8 for the basic characteristics test of viscoelastic springs in Section 9.4.3 – were applied to determine the basic characteristics of the spring oil dampers. For the evaluation of test results, data of the third sine wave cycle was sampled, while no smoothing (moving average calculation) was performed on the measured data. Table 9.17 describes the test conditions. In terms of the sampling frequency of measured data, it was adjusted so that the number of data per waveform would fall in the 400 to 500 range for each vibration cycle, so 100 Hz and 2 kHz were chosen. All measured data were processed through a 100-Hz low-pass filter.

Figures 9.26 through 9.30 provide the results of this sine wave vibration test, in the form of charts indicating the relationship between force and displacement related to the median cycle of the sine wave vibration. The acronyms "B.O.D.," "O.D.," and "S.U." shown in Figures and Tables 9.18 and 9.19 mean "spring oil damper," "oil damper," and "spring unit," respectively. Also, in each of the following Figures, "(a)" is the force generated by the entire spring oil damper while "(b)" is the force generated by the oil damper and "(c)" the force generated by the spring unit. In each figure where

Table 9.17 Conditions of the sine wave vibration test

Cycle [s]	Excitation amplitude [mm]					
	1	2.5	5	10	20	30
0.25	O	O	O	—	—	—
0.5	O	O	O	O	—	—
1	O	O	O	O	O	—
2	O	O	O	O	O	O
4	O	O	O	O	O	O

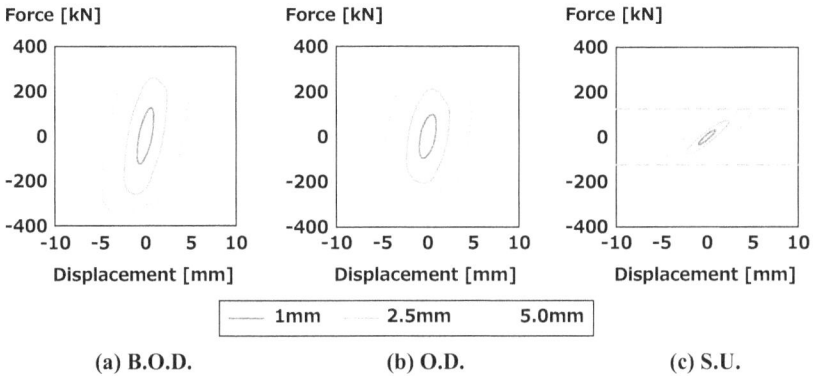

Figure 9.26 Chart indicating relation between force and displacement: Sine wave vibration test, cycle 0.25 sec.

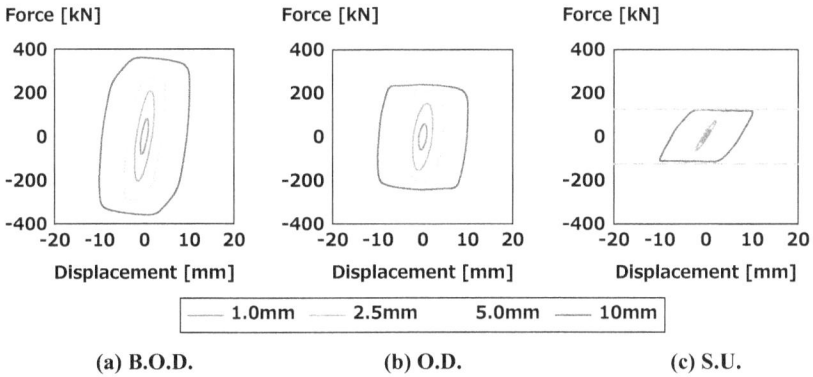

Figure 9.27 Chart indicating relation between force and displacement: Sine wave vibration test, cycle 0.5 sec.

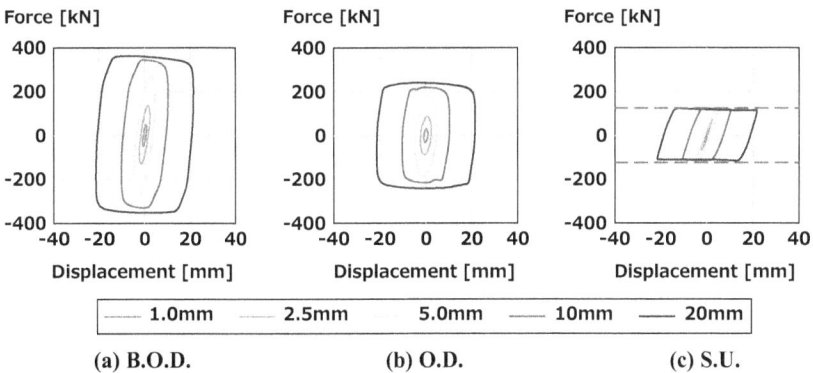

Figure 9.28 Chart indicating relation between force and displacement: Sine wave vibration test, cycle 1.0 second.

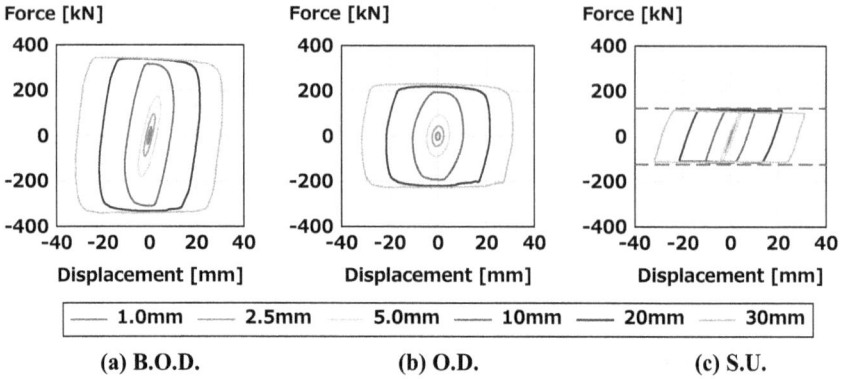

Figure 9.29 Chart indicating relation between force and displacement: Sine wave vibration test, cycle 2.0 sec.

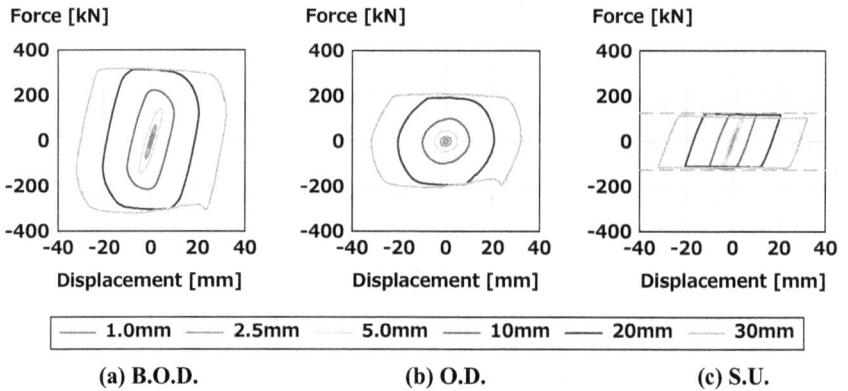

Figure 9.30 Chart indicating relation between force and displacement: Sine wave vibration test, cycle 4.0 sec.

"(c)" appears, the friction force of 125 kN at which the friction mechanism gets activated is represented by the dashed line. In addition, Figures 9.31 through 9.33 indicate the superimposed curves representing the measured data related to the spring oil damper as a whole and the oil damper while Table 9.18 provides a summary of data related to energy absorption by the spring oil damper and the oil damper, and the ratio of energy absorption by the oil damper to that by the spring oil damper.

Examination of the force-displacement charts in Figures 9.26 through 9.30 revealed that the oil damper exhibited such performance that would be represented by an oval in the primary attenuation coefficient phase and by a near parallelogram in the secondary attenuation coefficient phase in response to sine wave vibration, as the conventional oil damper would.

Table 9.18 Energy absorption capacity data: Sine wave vibration test

Cycle [s]	Excitation amplitude [mm]	O.D. [kN·m]	B.O.D. [kN·m]	O.D./B.O.D. [%]
0.25	1	0.249	0.283	88.0
	2.5	1.452	1.663	87.3
	5	3.918	4.892	80.1
0.5	1	0.153	0.172	89.1
	2.5	0.992	1.111	89.3
	5	3.114	3.711	83.9
	10	8.789	12.120	72.5
1	1	0.081	0.091	88.9
	2.5	0.620	0.702	88.2
	5	2.657	3.146	84.5
	10	8.026	11.332	70.8
	20	19.352	27.877	69.4
2	1	0.047	0.054	87.6
	2.5	0.330	0.388	85.1
	5	1.355	1.719	78.8
	10	6.475	9.655	67.1
	20	17.229	25.568	67.4
	30	27.006	39.584	68.2
4	1	0.028	0.032	86.5
	2.5	0.148	0.181	81.3
	5	0.697	1.040	67.0
	10	2.999	5.896	50.9
	20	12.862	20.657	62.3
	30	23.035	35.606	64.7

Table 9.19 Energy absorption capacity: Random wave vibration test

Excitation amplitude [mm]	O.D. [kN·m]	B.O.D. [kN·m]	O.D./B.O.D. [%]
5	20.588	23.732	86.8
20	215.407	269.743	79.9

As for the spring unit and its performance below and up to the excitation amplitude of 5 mm where the friction mechanism would get activated, with cycles of 1 sec. and 4 sec., the characteristic was almost linear. With a cycle of 0.25 sec., however, it exhibited performance represented as this oval which is typical of any viscoelastic material. When the excitation amplitude was raised past 5 mm, where the friction mechanism would get activated, the friction mechanism apparently exhibited its steady operation, manifesting the specified friction force of 125 kN, under all the test conditions regardless of vibration cycle or excitation amplitude. In Figures 9.31

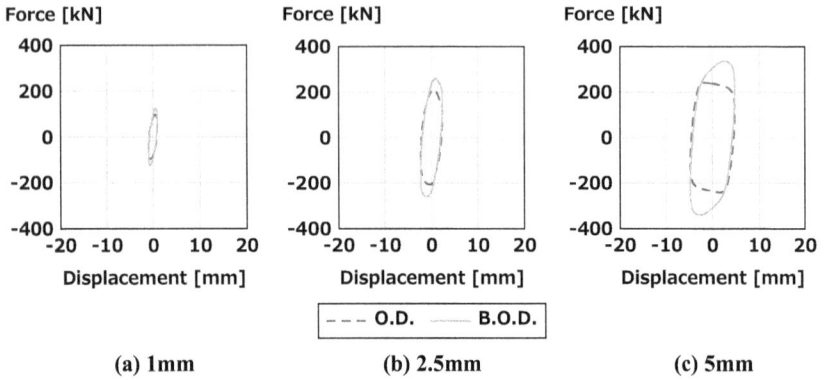

(a) 1mm   (b) 2.5mm   (c) 5mm

Figure 9.31 Chart indicating relation between force and displacement (superimposed):
Sine wave vibration test, cycle 0.25 sec.

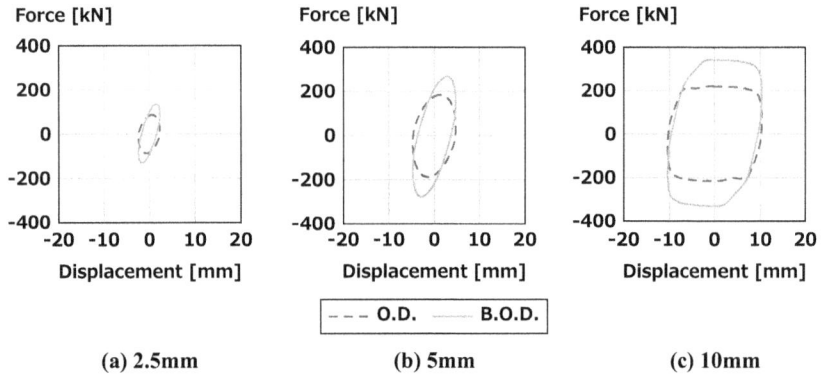

(a) 2.5mm   (b) 5mm   (c) 10mm

Figure 9.32 Chart indicating relation between force and displacement (superimposed):
Sine wave vibration test, cycle 1.0 second.

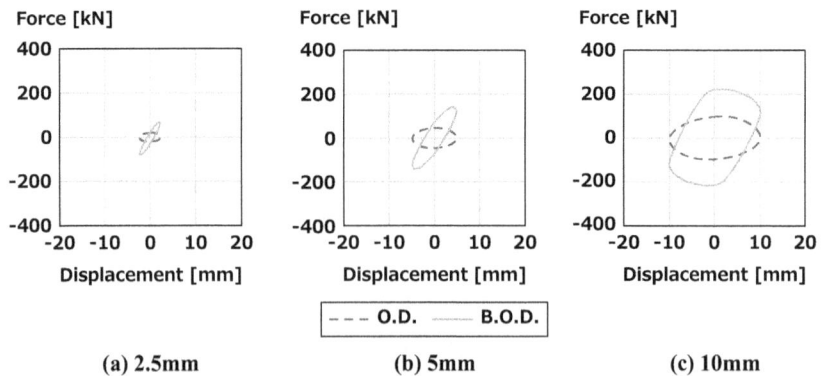

(a) 2.5mm   (b) 5mm   (c) 10mm

Figure 9.33 Chart indicating relation between force and displacement (superimposed):
Sine wave vibration test, cycle 4.0 sec.

through 9.33 indicating the superimposed curves of the measured data related to the spring oil damper as a whole and the oil damper, the performance curve of the spring oil damper exhibited a form similar to that of the oil damper, up to the excitation amplitude of 5 mm, except that the upward slope would be more pronounced due to the additional stiffness provided by the viscoelastic spring. When the excitation amplitude exceeded 5 mm, the curve of the spring oil damper would take a shape again similar to that of the oil damper but only stretched wider along the y-axis due to the friction force resulting from the friction mechanism. In Table 9.18, where the ratio of the oil damper's energy absorption to that of the entire spring oil damper is indicated, the ratio ranged between 70 and 90% up to the excitation amplitude of 5 mm, which was the trigger point for friction mechanism activation, while the ratio remained around 70% after the excitation amplitude had surpassed 5 mm and the friction mechanism had been activated. Based on the above, it was determined that up to the excitation amplitude of 5 mm, the viscoelastic spring absorbed roughly 10 to 30% of the entire energy absorbed by the spring oil damper, but once the excitation amplitude became greater than 5 mm, the viscoelastic spring in conjunction with the friction mechanism absorbed about 30 to 50% of the total energy absorbed by the spring oil damper.

9.5.4 Random wave vibration test

In this segment of the test, random vibrations mimicking actual seismic waves were applied to the spring oil dampers to determine if each of their components would operate smoothly as intended. For this test, the single degree-of-freedom model developed based on the actual factory building as described in Chapter 11 was used to analyze and simulate the relative story displacement waveforms that would arise at the point mass of BCJ-L1 and BCJ-L2 input, while adjusting the amplification as such that the maximum excitation amplitude would become 5 mm with BCJ-L1 and 20 mm with BCJ-L2. Figure 9.34 indicates the displacement waveforms resulting from the aforementioned seismic inputs.

Figures 9.35 and 9.36 provide the results of this random wave vibration test, in the form of charts indicating the relationship between force and displacement. In addition, Table 9.19 provides a summary of data related to energy absorption by the spring oil damper and the oil damper, and the ratio of energy absorption by the oil damper to that by the spring oil damper.

Examination of the force-displacement charts in Figures 9.35 and 9.36 revealed that the friction mechanism in the spring oil damper operated smoothly and stably also in response to the random wave inputs. According to Table 9.19, where the ratio of the oil damper's energy absorption to that of the entire spring oil damper is indicated, the ratio remained around 15%

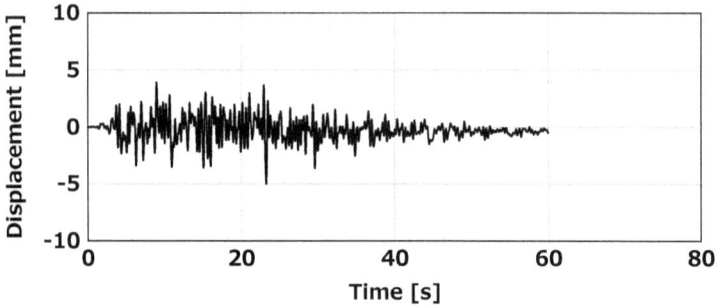

(a) BCJ-L1

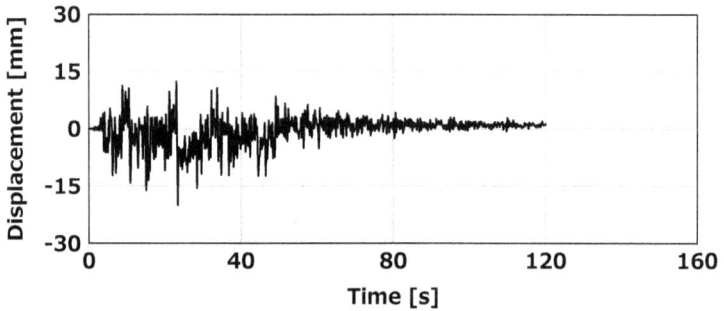

(b) BCJ-L2

Figure 9.34 Relative story displacement waveforms obtained from single degree-of-freedom model analysis.

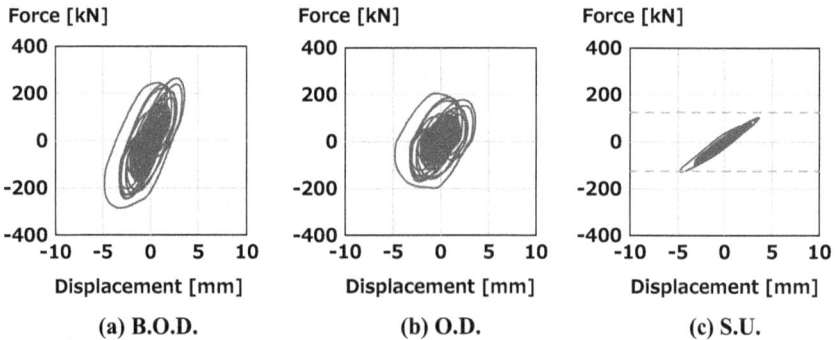

(a) B.O.D.

(b) O.D.

(c) S.U.

Figure 9.35 Chart indicating relation between force and displacement: Random wave vibration test, BCJ-L1, maximum excitation amplitude 5 mm.

Force [kN]

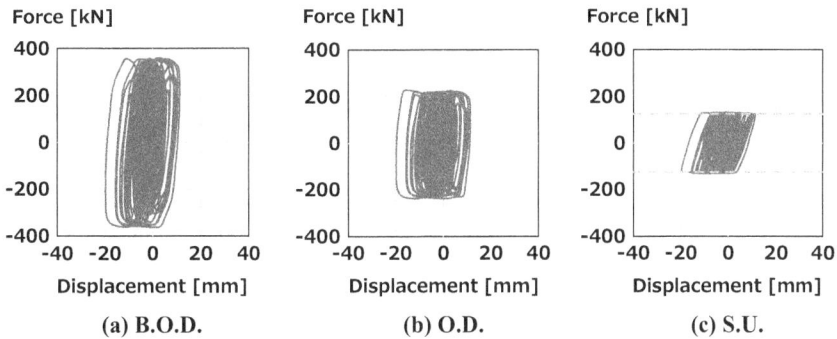

(a) B.O.D.

Force [kN]

(b) O.D.

Force [kN]

(c) S.U.

Figure 9.36 Chart indicating relation between force and displacement: Random wave vibration test, BCJ-L2, maximum excitation amplitude 20 mm.

up to the excitation amplitude of 5 mm, which was the threshold for activating the friction mechanism, while the ratio was around 20% at the excitation amplitude of 20 mm after the friction mechanism had been activated.

9.6 SUMMARY

In this chapter, the spring oil damper configuration was first explained, and then the results of their standalone performance tests were presented, based on which the basic characteristics of the spring oil dampers were determined and their multi-factor dependence and durability after having been put through the series of tests were observed.

First, Section 9.2 described the configuration of the spring oil dampers. It was explained that each spring oil damper had the oil damper placed in the middle, flanked by the unit comprised of the viscoelastic spring and the friction mechanism on each side, whose functions were vibration control and stiffness support.

Then, Section 9.3 provided examples of structural reinforcement using these spring oil dampers, while stating that the spring oil dampers would require less reinforcement points and leave more openings available for use after reinforcement than reinforcement by conventional braces and oil dampers would, in reinforcing low-stiffness structures such as steel-framed factories and warehouses based on outdated seismic resistance standards. It was also explained that less reinforcement points would not only reduce the cost of reinforcement components but also the cost of reinforcement related to columns, beams, footing beams, etc.

Section 9.4 focused on the testing of the spring oil dampers with oil drained, to determine the standalone performance of the spring units each comprised of the viscoelastic spring and the friction mechanism. By applying sine wave vibration of various cycles to the spring oil dampers and observing the results, it was determined that the spring units had exhibited performance close to their design-specified value and that their performance had not changed in any major way in response to change in excitation amplitude or velocity. In addition, the durability tests that involved repetitive vibration inputs simulating major earthquake events as well as normal winds and minor to medium-intensity earthquakes were described and it was determined that the spring units exhibited sufficient durability. Furthermore, in the random wave vibration test, random wave inputs mimicking actual earthquakes were applied, and the spring units were determined to have operated smoothly.

Finally, Section 9.5 described the performance tests that were conducted on the spring oil dampers, the performance of which had been adjusted to the test apparatus specifications, and the entire spring oil dampers' performance was observed. It was then determined that the amount of energy absorption resulting from seismic attenuation by the stiffness-supporting viscoelastic material was 10 to 30% of the total energy absorbed by the entire spring oil dampers in response to normal winds and minor to medium-intensity earthquakes.

REFERENCES

1. The Japan Society of Seismic Isolation (JSSI): *Passive Seismic Control Structure Design and Construction Manual*, 2nd Edition, July 2007. (in Japanese). JSSI.

Chapter 10

Analysis model for stiffness-supported oil dampers

10.1 INTRODUCTION

This chapter explains about analysis models that can be applied to spring oil dampers and compares their performance against the results of the tests described in Section 9.5. First, Section 10.2 describes several models of analyzing spring oil dampers. Section 10.3 then compares the test results presented in Section 9.5 with the results of the analysis performed using the analysis model being proposed in this dissertation, utilizing the displacement waveform data obtained during the aforementioned tests.

10.2 ANALYSIS MODEL FOR STIFFNESS-SUPPORTED OIL DAMPERS

This section describes several analysis models that can be applied to spring oil dampers. Section 10.2.1 explains how the Maxwell model was modified into a simple computation model in this study and how it can be applied to oil dampers, treating each viscoelastic spring as a linear spring and each friction mechanism as a friction element. Section 10.2.2 takes this simplified computation model and further modifies it into a version of the Voigt model as the improved analysis model being suggested in this study, which takes into account the effects of the viscous property of each viscoelastic spring used.

10.2.1 Simple computation model

In published papers,[1] a spring oil damper is usually described in such a graphical representation as indicated in Figure 10.1, using the Maxwell model and nonlinear springs (i.e., spring and friction models).

In published studies,[2] the computation method applied to a Maxwell model, where a spring and a dashpot possessing such bilinear attenuation characteristics as shown in Figure 10.2 are serially connected, usually follows the procedure as described below.

DOI: 10.1201/9781003290261-12

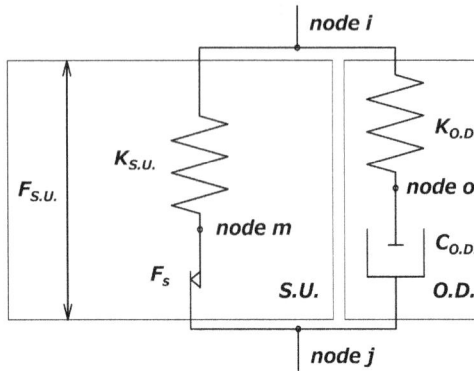

Figure 10.1 Spring oil damper analysis model: Simple computation model.

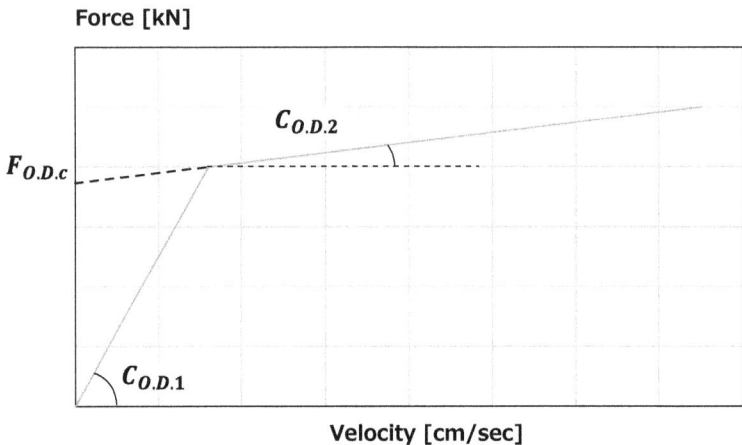

Figure 10.2 Chart indicating relation between force and velocity of dashpots.

The salient advantage of this computation method is its simplicity, as it can be applied to any instance where the amount of data required for numerical calculation is limited and the unit time Δt of the analysis period is sufficiently small, to approximate the solution by dividing the change in displacement per time interval in relation to the velocity specific to each time t. Research indicates that this analysis technique provides sufficient precision as long as the time interval does not exceed 0.01 sec.[2]

The procedure for calculating the force involving a set of Maxwell elements is explained below. First, the balance of force at the connection point O must be determined. In the formula below, u = relative displacement and v = relative velocity. The force involving the Maxwell elements $F_{kO.D.}$ is expressed by the formula below, where: $F_{cO.D.}$ = the spring's elastic force; and $F_{O.D.ij}$ = the dashpot's viscous force.

$$F_{k\,O.D.} = F_{c\,O.D.} = F_{O.D.ij} \tag{10.1}$$

The spring force F_k is calculated by the formula below, where: u_{ij} = relative displacement among the Maxwell elements; and u_{oj} = relative displacement between the dashpots.

$$F_{k\,O.D.} = K_{O.D.}\left(u_{ij} - u_{oj}\right) \tag{10.2}$$

The force generated by the dashpots following relief is expressed by the formula below, where: $C_{O.D.2}$ = second gradient; and $F_{O.D.c}$ = force at the point of intersection with the force axis.

$$F_{c\,O.D.} = C_{O.D.2}v_{oj} + F_{O.D.c} \tag{10.3}$$

Hence, the balance condition at the connection point O can be calculated as follows, plugging (10.2) and (10.3) into the formula (10.1).

$$K_{O.D.}\left(u_{ij} - u_{oj}\right) = C_{O.D.2}v_{oj} + F_{O.D.c} \tag{10.4}$$

If the unit time Δt is small enough, the velocity at a given time t can be approximated by the average velocity from $t - \Delta t$ to t, so the velocity at the time t can be estimated by the following formula.

$${}^{t}v_{oj} = \frac{\Delta^{t}u_{oj}}{\Delta t} \tag{10.5}$$

Here, ${}^{t}v_{oj}$ is the relative velocity between the dashpots at the time t, whereas $\Delta^{t}u_{oj}$ is the increase in relative displacement between the dashpots at the time t. Plugging the (10.5) into the formula (10.4) yields the following formula.

$$K_{O.D.}({}^{t}u_{ij} - {}^{t}u_{oj}) = C_{O.D.2}\left(\frac{\Delta^{t}u_{oj}}{\Delta t}\right) + F_{O.D.c} \tag{10.6}$$

In addition, the relative displacement at the time t can be expressed by the sum of the time $t - \Delta t$ and the increase in relative displacement as follows.

$${}^{t}u_{oj} = {}^{t-\Delta t}u_{oj} + \Delta^{t}u_{oj} \tag{10.7}$$

Applying the formula (10.7) to rearrange the formula (10.6) as such that it is made about $\Delta^{t}u_{oj}$ yields the following formula.

$$\Delta^{t}u_{oj} = \frac{K_{O.D.}({}^{t}u_{ij} - {}^{t-\Delta t}u_{oj}) - F_{O.D.c}}{\dfrac{C_{O.D.2}}{\Delta t} + K_{O.D.}} \tag{10.8}$$

Then, plugging (10.8) into the formula (10.7) to determine $\Delta^t u_{oj}$ and then plugging it into the formula (10.2) yields the spring force as well as the Maxwell element force.

Next, Figure 10.3 provides a flowchart explaining the numerical calculation algorithm that allows for the computation of the force generated by a spring oil damper undergoing displacement, using the conventional model.

First, focusing on the oil damper part, $\Delta^t u_{oj}$ (amount of increase in displacement between the dashpots in the linear range, before relief) is calculated based on $^t u_{ij}$ (relative displacement between the dampers) and $^{t-\Delta t} u_{oj}$ (relative displacement between the dashpots) that were obtained in the previous step. This can be achieved using the formula below, which is a modification of the formula (10.8) where $C_{O.D.2}$ is replaced by $C_{O.D.1}$ and $F_{O.D.c}$ by 0.

$$\Delta^t u_{oj} = \frac{K_{O.D.}(^t u_{ij} - {}^{t-\Delta t} u_{oj})}{\dfrac{C_{O.D.1}}{\Delta t} + K_{O.D.}} \tag{10.9}$$

Next, v_{oj} (relative velocity between the dashpots) is calculated by plugging (10.9) into the formula (10.5), to make a determination on the relief. If this velocity turns out to be higher than the relief velocity on the positive side, the amount of increase in relative displacement between the dashpots is recalculated using the formula (10.8). Then, if it turns out to be lower than the relief velocity on the negative side, another calculation is performed using the formula below, which is a modification of the formula (10.8) where $F_{O.D.c}$ is replaced by $-F_{O.D.c}$.

$$\Delta^t u_{oj} = \frac{K_{O.D.}(^t u_{ij} - {}^{t-\Delta t} u_{oj}) + F_{O.D.c}}{\dfrac{C_{O.D.2}}{\Delta t} + K_{O.D.}} \tag{10.10}$$

This $\Delta^t u_{oj}$ is plugged into the formula (10.7) to calculate u_{oj} (relative displacement between the dashpots) and u_{oi} (relative displacement between the springs). Then, u_{oi} is multiplied by $K_{O.D.}$ (oil damper stiffness) to obtain the force generated by the oil damper.

As for the spring unit, its performance in terms of force-displacement relation is expressed as a linear line, up to the point of friction mechanism activation. After the friction force initiating the friction mechanism's operation reaches F_s, the displacement continuously increases as the force remains constant. Hence, the force generated by the spring unit can be calculated by the formula below.

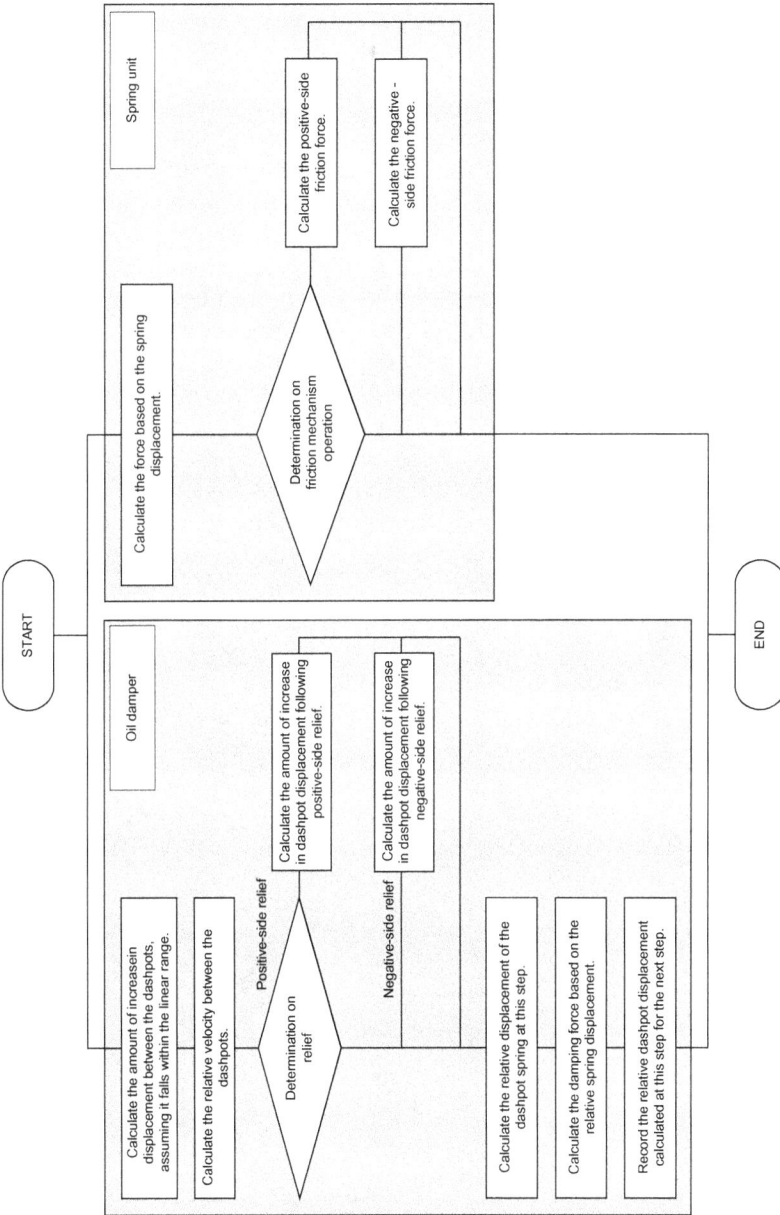

Figure 10.3 Flowchart for calculating numerical data related to spring oil dampers: Simple computation model.

$$F_{bij} = \begin{cases} K_{S.U.} \cdot u_{ij} & \left| K_{S.U.} \cdot u_{im} + C_{S.U.} \cdot \dot{u}_{im} \right| \leq F_s \\ F_s sgn\left(\dot{u}_{mj}\right) & \left| K_{S.U.} \cdot u_{im} + C_{S.U.} \cdot \dot{u}_{im} \right| \geq F_s \end{cases}$$

(10.11)

10.2.2 Improved model

Examination of the spring oil damper test results provided in Section 9.5 revealed that, as long as the excitation amplitude did not exceed 5 mm, the viscoelastic spring unit had absorbed roughly 10 to 30% of all energy absorbed by the entire spring oil damper. Therefore, the viscoelastic spring analysis method was switched to the Voigt model from the linear spring model to examine how the viscous property of the viscoelastic material used in the spring unit would manifest its effects. Figure 10.4 indicates a diagram of this improved model applied to the spring oil damper analysis.

As indicated in Figure 10.4, this spring unit analysis method takes the form of a hybrid model combining the Voigt and friction models. Hence, as with the computation technique applied to oil dampers as described earlier, the force generated by the dashpots can be calculated based on the velocity between nodes i and j indicated above and adding to it the spring-generated force. Then, if the sum turns out to be greater than the friction force, a constant force can be maintained based on the friction model. Hence, the force generated by the spring unit can be calculated as shown below. In addition, Figure 10.5 provides a flowchart explaining the numerical calculation procedure.

$$F_{S.U.} = \begin{cases} K_{S.U.} \cdot u_{ij} + C_{S.U.} \cdot \dot{u}_{oj} & \left| K_{S.U.} \cdot u_{im} + C_{S.U.} \cdot \dot{u}_{im} \right| \leq F_s \\ F_s sgn\left(\dot{u}_{pj}\right) & \left| K_{S.U.} \cdot u_{im} + C_{S.U.} \cdot \dot{u}_{im} \right| \geq F_s \end{cases}$$

(10.12)

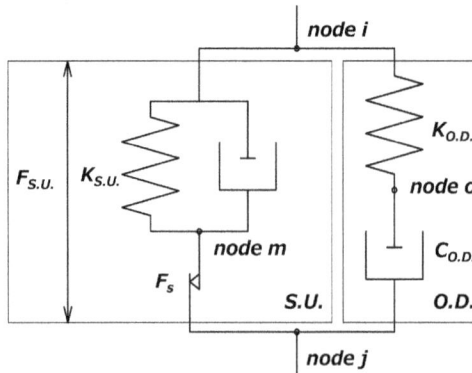

Figure 10.4 Analysis model for spring oil dampers: Improved model.

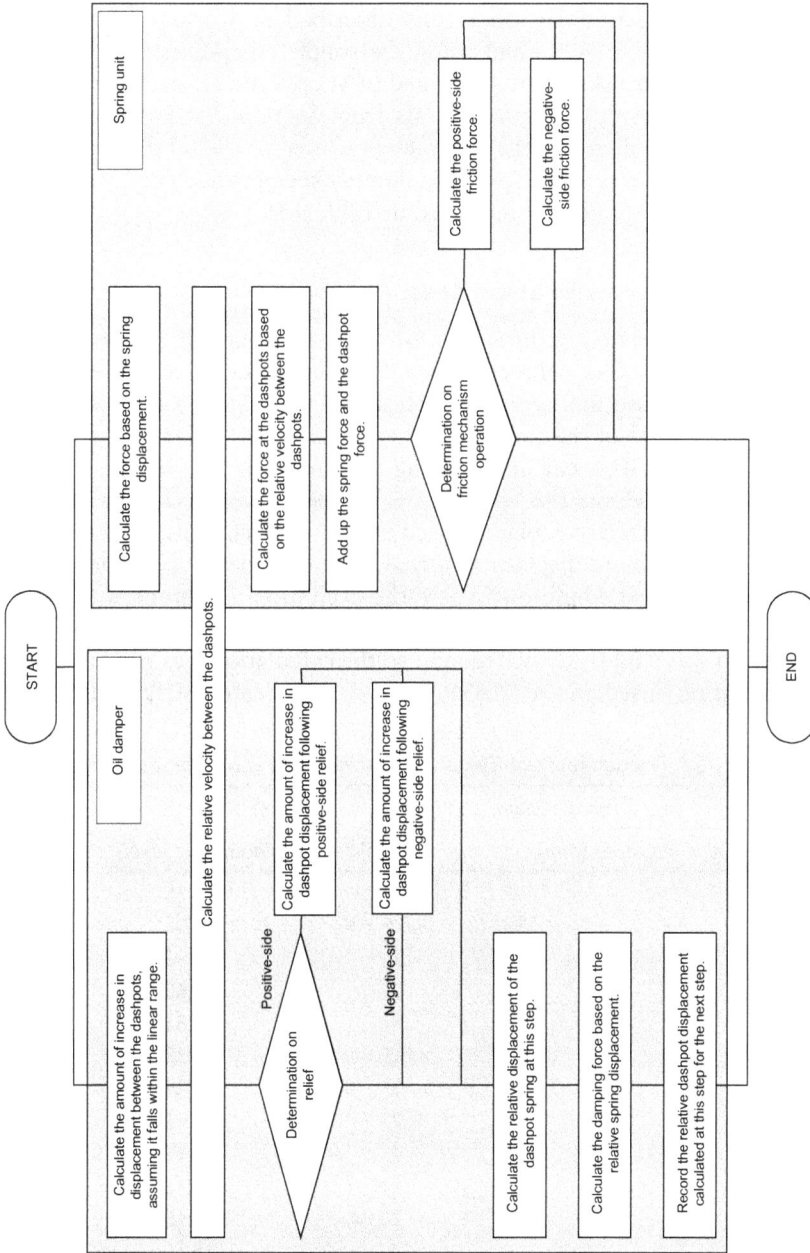

Figure 10.5 Flowchart for calculating numerical data related to spring oil dampers: Improved model.

10.3 COMPARISON BASED ON PERFORMANCE TEST OF STIFFNESS-SUPPORTED OIL DAMPERS

This section compares the test results described in Section 9.5 with the results of the analysis obtained using the simple computation model and the improved model that were explained in Section 10.2, based on the displacement data resulting from the tests from Section 9.5. For the following numerical calculations, the parameters related to the oil dampers were taken from the figures in the spring oil damper specifications that were used in the tests in Section 9.5, as indicated in Table 9.16.

10.3.1 Sine wave vibration test

The parameters related to the spring unit were calculated based on the test results using the method set forth below. For the friction force of the friction element, the friction mechanism's average force measured while in operation under each set of the test conditions was applied. Table 10.1 provides the friction force data calculated using this method. The stiffness of the linear spring model and the Voigt model related to each viscoelastic spring of the spring units was calculated based on the maximum displacement and the force at maximum displacement that were obtained during the related tests, under the test conditions where the excitation amplitude was 1 mm and 2.5 mm. As for when the excitation amplitude was 5 mm, the friction mechanism was slightly activated, and so the calculation was performed by approximation based on the friction force data indicated in Table 10.1 and

Table 10.1 Friction force of friction mechanism: Sine wave vibration test

Test conditions		
Excitation amplitude [mm]	Cycle [s]	Friction force [kN]
5	0.25	113.75
	0.5	116.06
	1	117.33
	2	119.68
	4	119.53
10	0.5	115.32
	1	116.84
	2	117.81
	4	117.91
20	1	114.16
	2	114.15
	4	115.11
30	2	110.16
	4	110.26

the viscoelastic springs' force-displacement curves obtained in the preliminary test as indicated in Figure 9.20. Because any viscoelastic spring has such characteristic that causes its stiffness to change due to its dependence on displacement, the stiffness was calculated based on the results of the test where the excitation amplitude was 5 mm. For each instance where the excitation amplitude was greater than 5 mm, the stiffness from the test conditions where the excitation was 5 mm for each cycle was applied. For the attenuation coefficient relating to the dashpots in the Voigt model, under the test conditions where the friction mechanism was inactive, the equivalent attenuation coefficient calculated from the hysteretic area obtained from the test results was applied, as expressed in the formula (10.13) below.

$$C_{S.U.} = \frac{\Delta W}{\pi \cdot u_{max}^2 \cdot \omega} \tag{10.13}$$

where: $C_{S.U.}$ = spring unit's equivalent viscous attenuation coefficient [kNs/mm]; ΔW = energy consumption per cycle [kN·mm]; u_{max} = maximum excitation amplitude [mm]; and ω = circular frequency (= $2\pi f$) [rad/s].

Under several test conditions where the excitation amplitude was 5 mm, the friction mechanism got activated, so the equivalent viscous attenuation coefficients were applied that were calculated assuming that the equivalent viscous attenuation coefficients calculated from the test results – where the excitation amplitude was 1 mm and 2.5 mm in each cycle – would change linearly. For the rest of the test conditions where the excitation amplitude was greater than 5 mm, the equivalent viscous attenuation coefficients estimated at the excitation amplitude of 5 mm were applied. Table 10.2 provides the stiffness data and the equivalent viscous attenuation coefficients for each cycle, calculated using the method as explained above for the sine wave vibration test. Examination of Table 10.2 revealed that, due to the viscoelastic springs' dependence on displacement, the stiffness had decreased as the excitation amplitude had increased. The parenthesized figures in Table 10.2 are estimated equivalent viscous attenuation coefficients at the excitation amplitude of 5 mm, inferred from the test results where the excitation amplitude was 1 mm and 2.5 mm.

Figures 10.6 through 10.12 provide superimposed graphs of the test and analysis results. In addition, Table 10.3 presents data on energy absorption by the spring unit per sine wave cycle, which was calculated based on the test results and the analyses drawn using each of the analysis models, so that the energy absorption of the viscoelastic springs across the different models could be compared. Then, Figure 10.4 indicates the rates of deviation between the test results and the analysis results relating to Table 10.3.

Examination of the graphs indicating the force-displacement relations in Figures 10.6 through 10.12 revealed that the analysis results accurately tracked the actual test results related to the oil damper.

Table 10.2 Viscoelastic spring stiffness and attenuation coefficient: Sine wave vibration test

	Test conditions		
Cycle [s]	Excitation amplitude [mm]	Stiffness [kN/mm]	Attenuation coefficient [kNs/mm]
0.25	1	32.71	0.50
	2.5	29.89	0.48
	5	25.87	(0.43)
0.5	1	32.40	0.56
	2.5	30.01	0.61
	5	25.77	(0.68)
1	1	32.03	0.63
	2.5	29.62	0.75
	5	25.71	(0.93)
2	1	31.78	0.80
	2.5	29.28	0.99
	5	25.61	(1.32)
4	1	31.88	1.03
	2.5	29.66	1.33
	5	25.62	(1.85)

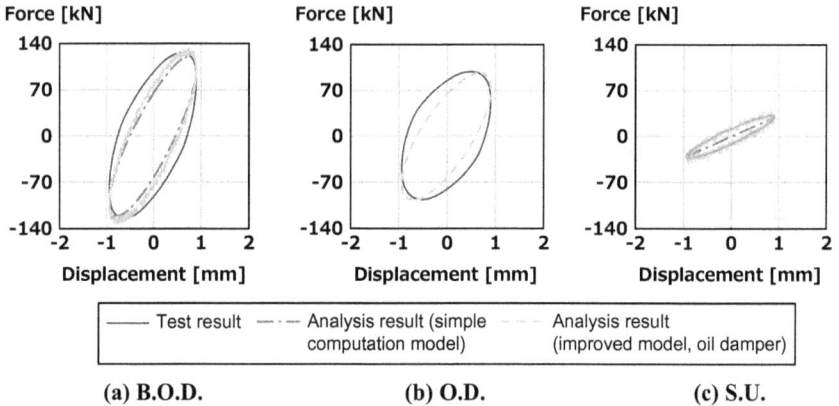

——— Test result	—·— Analysis result (simple computation model)	- - - Analysis result (improved model, oil damper)

(a) B.O.D.   **(b) O.D.**   **(c) S.U.**

Figure 10.6 Chart indicating relation between force and displacement (superimposed analysis): Sine wave vibration test, cycle 0.25 sec., excitation amplitude 1 mm.

Examination of Tables 10.3 and 10.4 revealed that the improved model had performed better than the simple computation model, in terms of how closely they tracked the actual test results expressed in the energy absorption breakdown and the deviation rates, across all test conditions. Indeed, the improved model was able to achieve an accuracy of roughly ±10%.

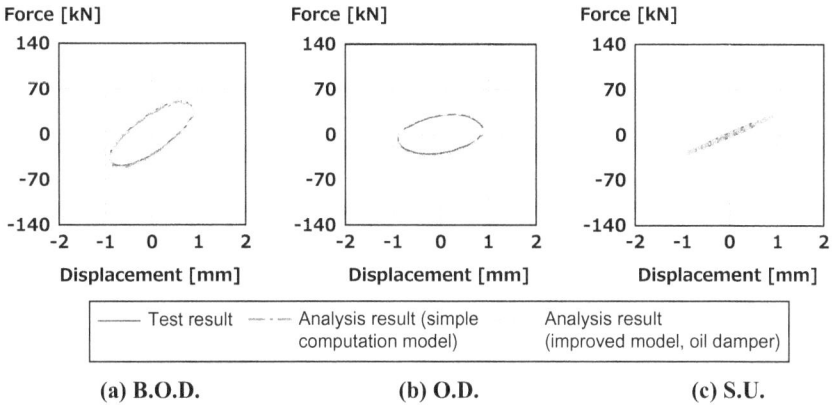

Figure 10.7 Chart indicating relation between force and displacement (superimposed analysis): Sine wave vibration test, cycle 1 sec., excitation amplitude 1 mm.

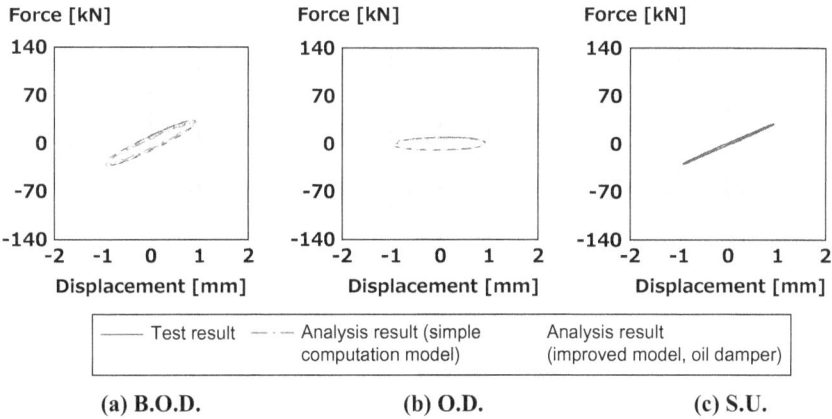

Figure 10.8 Chart indicating relation between force and displacement (superimposed analysis): Sine wave vibration test, cycle 4 sec., excitation amplitude 1 mm.

Based on the aforementioned comparisons, it was determined that it would be better to apply the Voigt model instead of the linear spring model in estimating the results of tests involving the viscoelastic springs in such spring oil damper configuration, as the Voigt model could take into account the attenuation characteristics of the viscoelastic springs and provide higher accuracy than the linear spring model could. It must be noted that the deviation was greater than 10% only under the test condition where the cycle was 4 sec. and the excitation amplitude was 5 mm. This might have been

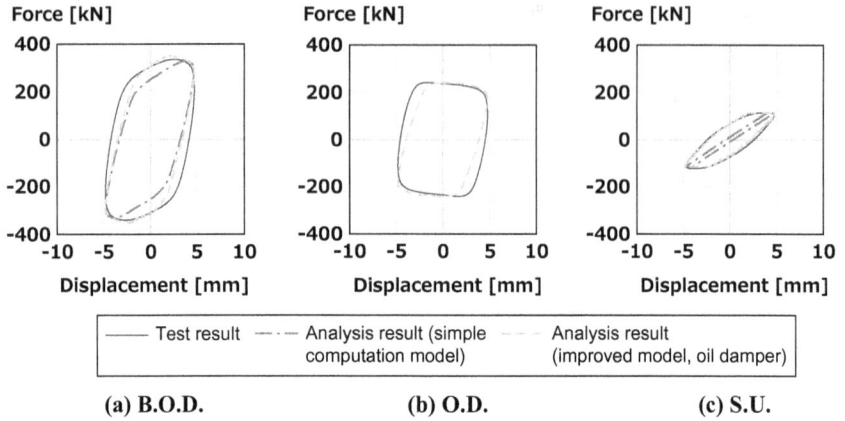

Figure 10.9 Chart indicating relation between force and displacement (superimposed analysis): Sine wave vibration test, cycle 0.25 sec., excitation amplitude 5 mm.

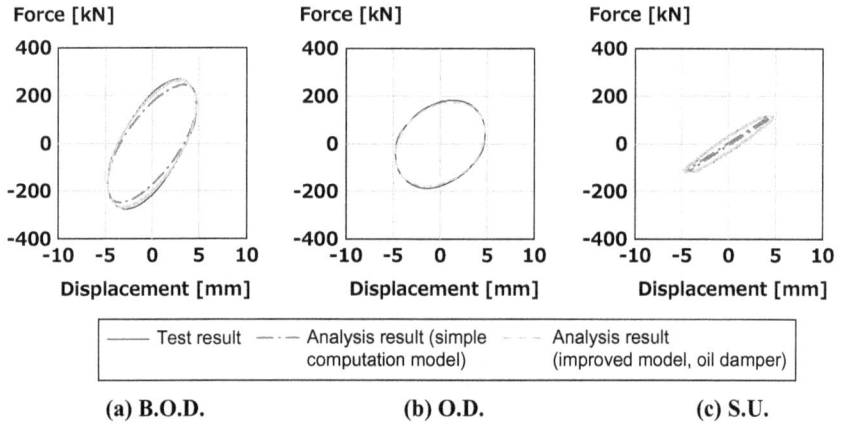

Figure 10.10 Chart indicating relation between force and displacement (superimposed analysis): Sine wave vibration test, cycle 1 sec., excitation amplitude 5 mm.

caused because the test results showed greater energy absorption than the analysis results did at such low excitation amplitude due to the viscoelastic spring's dependence on displacement.

10.3.2 Random wave vibration test

As the random wave vibration test was conducted after an intervening period following the sine wave vibration test, the test configuration had to be reassembled resulting in different friction mechanism settings. Therefore, after the aforementioned reassembly, another sine wave vibration test with a cycle

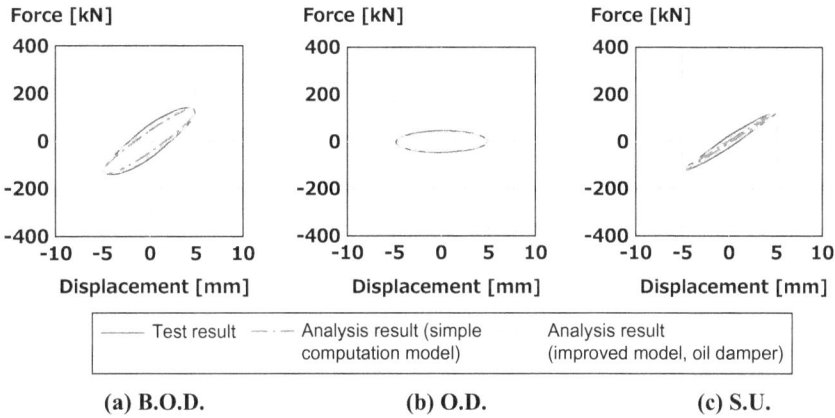

Figure 10.11 Chart indicating relation between force and displacement (superimposed analysis): Sine wave vibration test, cycle 4 sec., excitation amplitude 5 mm.

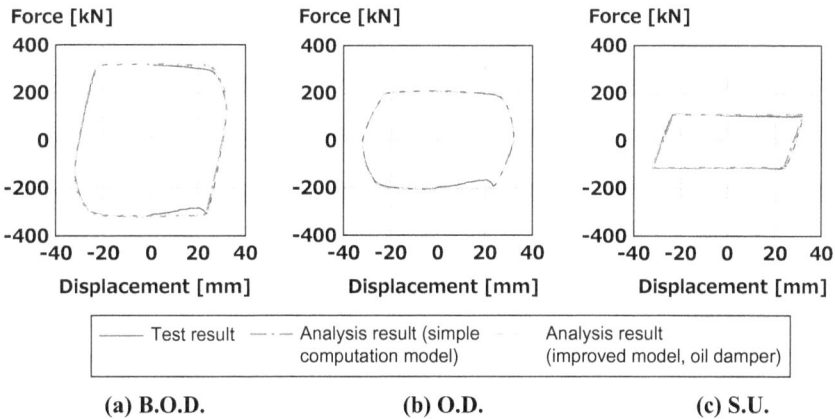

Figure 10.12 Chart indicating relation between force and displacement (superimposed analysis): Sine wave vibration test, cycle 4 sec., excitation amplitude 30 mm.

of 0.4 sec. (which is roughly equivalent to the cycle applied to the BODV analysis model for reinforcement using spring oil dampers as described in Chapter 11) and excitation amplitude of 5 mm and 20 mm was conducted, to calculate the friction mechanism's friction force and the viscoelastic spring's stiffness. As for the attenuation coefficient that would be applied to the Voigt model, the following calculation method was used. First, the data obtained from the sine wave vibration test related to the excitation amplitude of 1 mm and 2.5 mm for each cycle and the estimated data related to the excitation amplitude of 5 mm were considered together to find their average, which was then used to identify an approximation formula for each

Table 10.3 Energy absorption capacity of spring units: Sine wave vibration test

Test conditions		Energy absorption [kN·m]		
			Analysis results	
Cycle [s]	Excitation amplitude [mm]	Test results	Simple computation model	Improved model
0.25	1	0.034	0.000	0.036
	2.5	0.212	0.000	0.210
	5	0.975	0.179	0.891
0.5	1	0.019	0.000	0.020
	2.5	0.119	0.000	0.120
	5	0.597	0.038	0.612
	10	3.332	2.581	3.494
1	1	0.010	0.000	0.011
	2.5	0.083	0.000	0.083
	5	0.488	0.085	0.480
	10	3.306	2.734	3.435
	20	8.525	7.809	8.742
2	1	0.007	0.000	0.007
	2.5	0.058	0.000	0.058
	5	0.364	0.061	0.379
	10	3.180	2.785	3.298
	20	8.338	7.832	8.549
	30	12.579	11.980	12.776
4	1	0.004	0.000	0.005
	2.5	0.034	0.000	0.034
	5	0.343	0.097	0.289
	10	2.898	2.565	2.933
	20	7.795	7.428	7.950
	30	12.572	12.244	12.836

vibration frequency. Using this formula, the attenuation coefficient of the viscoelastic spring was calculated for a cycle of 0.4 sec. Figure 10.13 indicates the relationship between vibration frequency and attenuation coefficient. The friction mechanism's friction force in the random wave vibration test was calculated using the method as explained above, which was 122.07 kN for an excitation amplitude of 5 mm and 121.74 kN for an excitation amplitude of 20 mm. The stiffness in the linear spring and Voigt models was 25.51 kN/mm, while the viscous attenuation coefficient in the Voigt model was 0.56 kNs/mm. Figures 10.14 and 10.15 provide superimposed graphs of the test and analysis results. In addition, Table 10.5 presents energy absorption data under random wave vibration, which was calculated based on the spring unit test results and the analyses drawn using each of the analysis

Table 10.4 Deviation rates in energy absorption capacity: Sine wave vibration test

Test conditions		Rates of deviation in energy absorption between analysis results and test results [%]	
Cycle [s]	Excitation amplitude [mm]	(Simple computation model-test results) / Test results	(Improved model-test results) / Test results
0.25	1	−100.00	4.89
	2.5	−100.00	−1.00
	5	−81.68	−8.64
0.5	1	−100.00	4.71
	2.5	−100.00	0.76
	5	−93.56	2.44
	10	−22.55	4.86
1	1	−100.00	6.23
	2.5	−100.00	0.76
	5	−82.59	−1.54
	10	−17.30	3.90
	20	−8.40	2.54
2	1	−100.00	4.17
	2.5	−100.00	0.52
	5	−83.16	4.25
	10	−12.43	3.70
	20	−6.08	2.53
	30	−4.77	1.56
4	1	−100.00	5.77
	2.5	−100.00	1.54
	5	−71.62	−15.84
	10	−11.49	1.22
	20	−4.70	1.99
	30	−2.61	2.10

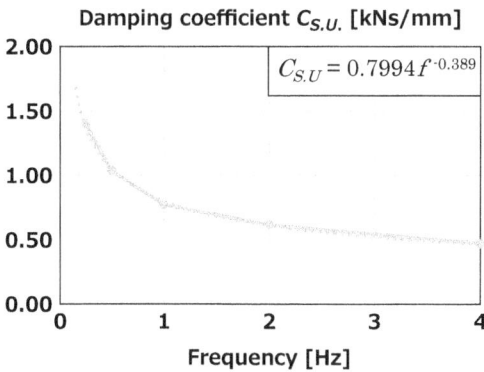

Damping coefficient $C_{S.U.}$ [kNs/mm]

$$C_{S.U} = 0.7994 f^{-0.389}$$

Figure 10.13 Chart indicating relation between attenuation coefficient and vibration frequency.

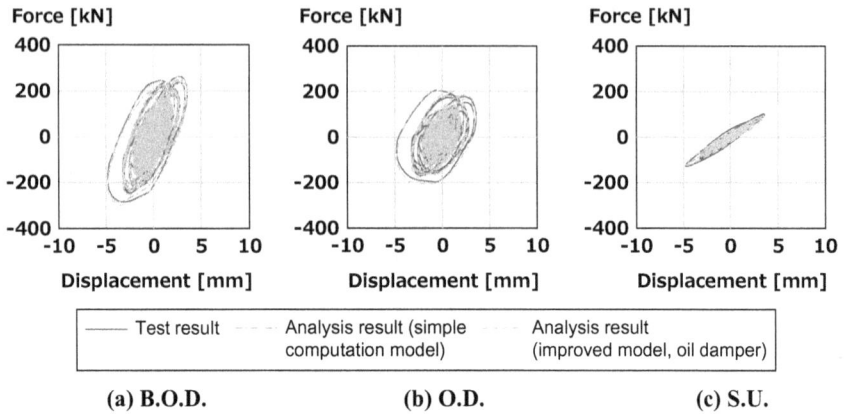

—— Test result	- - - Analysis result (simple computation model)	····· Analysis result (improved model, oil damper)

(a) B.O.D.   **(b) O.D.**   **(c) S.U.**

Figure 10.14 Chart indicating relation between force and displacement (superimposed analysis): Random wave vibration test, BCJ-L1, maximum excitation 5 mm.

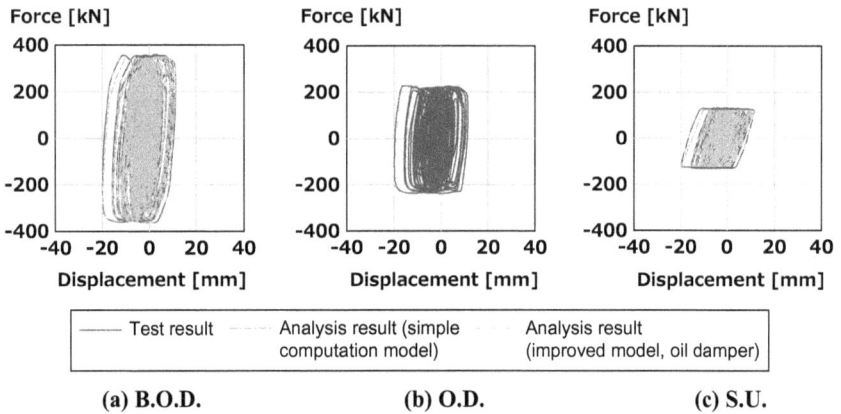

—— Test result	- - - Analysis result (simple computation model)	····· Analysis result (improved model, oil damper)

(a) B.O.D.   **(b) O.D.**   **(c) S.U.**

Figure 10.15 Chart indicating relation between force and displacement (superimposed analysis): Random wave vibration test, BCJ-L2, maximum excitation 20 mm.

Table 10.5 Energy absorption capacity of spring units: Random wave vibration test

Test conditions			Energy absorption [kN·m]	
				Analysis results
Waveform	Excitation amplitude [mm]	Test results	Simple computation model	Improved model
BCJ-L1	5	3.144	0.003	3.587
BCJ-L2	20	54.335	26.813	54.028

Table 10.6 Deviation rates in energy absorption capacity: Random wave vibration test

		Rates of deviation in energy absorption between analysis results and test results [%]	
Test results			
Waveform	Excitation amplitude [mm]	(Simple computation model-test results) / Test results	(Improved model-test results) / Test results
BCJ-L1	5	−99.90	14.09
BCJ-L2	20	−50.65	−0.57

models. Figure 10.6 indicates the rates of deviation between the test results and the analysis results relating to Table 10.5.

Examination of the graphs indicating the force-displacement relations in Figures 10.14 and 10.15 revealed that the analysis results accurately tracked the actual test results related to the oil damper. Furthermore, examination of Tables 10.5 and 10.6 revealed that the improved model had performed better than the simple computation model, in terms of how closely they tracked the actual test results expressed in the energy absorption break-down and the deviation rates, across all test conditions. It must be noted that the deviation was greater than 10% only under the test condition where the BCJ-L1 waveform was applied and the excitation amplitude was 5 mm. This must have been caused because the absolute amount of force involved might have been so small and prone to be affected by noise, and the noise-induced disturbance to the displacement waveforms seen in the test results might have impacted dashpot velocity approximation for the numerical calculation.

10.4 SUMMARY

In this chapter, the two types of analysis models for spring oil dampers were explained involving different modeling of viscoelastic springs, and the validity of the analysis model proposed in this study was verified by comparing the analysis results yielded by these models with the actual test results.

First, Section 10.2 described the numerical calculation processes involving the simple computation model and the improved model for spring oil damper analysis, where the former expressed the viscoelastic spring as a spring model while the latter expressed it as a Voigt model.

Then, Section 10.3 compared the spring oil damper test results as described in Section 9.5 with the results of the analysis obtained through the simple computation model and the improved model, applying the displace-ment waveform data from the test results. It was then determined that the improved model had performed better than the simple computation model, in terms of how closely they tracked the actual test results expressed in the energy absorption breakdown and the deviation rates, across all test conditions, achieving an accuracy of roughly ±10%. This comparative analysis revealed that it would be better to apply the Voigt model instead of the linear spring model in estimating the results of tests involving the viscoelastic springs in such spring oil damper configuration, as the Voigt model could take into account the attenuation characteristics of the viscoelastic springs and provide higher accuracy than the linear spring model could.

REFERENCES

1. A Yokoyama, O Takahashi, Y Asano: "Development of new oil damper with spring for architectural vibration control and experimental research on structural characterization", *Summaries of Scientific Lectures from the Architecture Institute of Japan (AIJ) Annual Meetings, Structure II*, pp.145–146, July 2016. (in Japanese)
2. O Takahashi, Y Tsuyuki, N Ikahata, Y Matsuzaki, T Fujita: "Development of oil damper and research on analytical model with consideration of damping characteristics", *Journal of Structural and Construction Engineering*, Architecture Institute of Japan (AIJ), No. 594, pp.49–56, Aug. 2005. (in Japanese)

Chapter 11

Examination involving time history response analysis based on single degree-of-freedom model

11.1 INTRODUCTION

This chapter aims to verify the efficacy of the spring oil dampers through time history response analysis using a single degree-of-freedom model of an actual factory building. First, Section 11.2 sets forth the analysis model specifications and analysis conditions. Then, Section 11.3 explains the analysis results and demonstrates the efficacy of the spring oil dampers.

11.2 ANALYSIS CONDITIONS AND ANALYSIS MODEL SPECIFICATIONS

For this analysis, a single degree-of-freedom model was used, representing an actual single-story factory building with a floor area of 8,960m². The modeling process only involved the factory building in the longitudinal direction, and SNAP-LE Ver.7 was the software used for the analysis. In terms of the computation method for this time history response analysis, the Newmark-β method (β=0.25) was applied, while the structural attenuation of the factory building's main structure was assumed to manifest as momentary stiffness proportional damping (h = 2%). Table 11.1 provides the specifications of this single degree-of-freedom model. As for the restoration force characteristics of the single degree-of-freedom model, a standard trilinear model was applied. Hence, stiffness reduction rates 1 and 2 as shown in Table 11.1 are the reduction rates of the secondary and tertiary stiffness relative to the primary stiffness in the trilinear model. For the time history response analysis, five different acceleration waveforms were used. Table 11.2 lists the names of the acceleration waveforms, their maximum velocity, and maximum acceleration. For each observed wave data besides BCJ-L1 and L2, maximum velocities of 25 cm/s and 50 cm/s were uniformly applied, while the original waveform models of BCJ-L1 and L2 were used in Lv. 1 and Lv. 2 analysis cases to transmit the acceleration waveforms to the structure's foundation accordingly. Figures 11.1 through 11.5 provide graphs indicating the acceleration-time relation specific to the

DOI: 10.1201/9781003290261-13

Table 11.1 Single degree-of-freedom model specifications

Mass M_b [t]	1343.22
Height H_b [cm]	610
Stiffness K_b [kN/cm]	383.3
Yielding load 1 [kN]	2465.82
Relative story drift angle at the first yield point [rad]	1/95
Stiffness reduction rate 1	0.714
Yielding load 2 [kN]	3137.97
Relative story drift angle at the second yield point [rad]	1/69
Stiffness reduction rate 2	0.031

Table 11.2 List of input waveforms

Type	Name	Maximum acceleration [cm/s²]	Maximum velocity [cm/s]
Actual observed waves	El Centro NS (1940)	341.7	33.66
	Taft EW (1952)	175.9	17.69
	Hachinohe NS (1968)	225.0	31.44
Simulated seismic waves	BCJ-L1 (Artificial)	207.3	29.10
	BCJ-L2 (Artificial)	355.7	57.40

Acceleration [cm/s²]

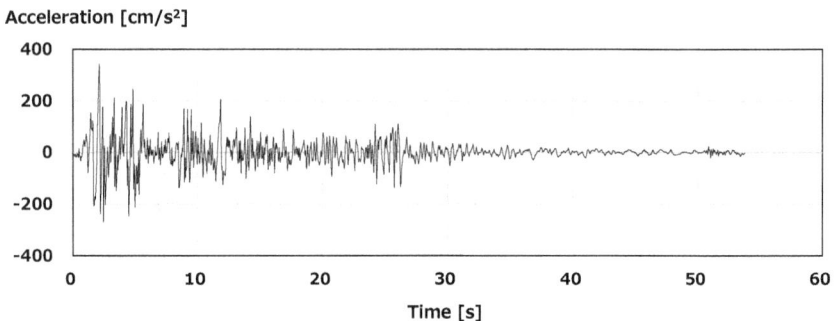

Figure 11.1 Chart indicating relation between acceleration and time: El Centro NS (1940).

acceleration waveforms. As for the types of analysis models employed, there were a total of three reinforcement models used, in addition to the model without any reinforcement (this analysis model is abbreviated "NO") as indicated in Figure 11.6, as follows: the model where six oil dampers and six steel braces were installed in 12 different locations in total (this analysis model is abbreviated "ODS"); the model where six spring oil dampers (modeled using the simple computation model) were installed in six different locations (this analysis model is abbreviated "BODS"); and the

Acceleration [cm/s²]

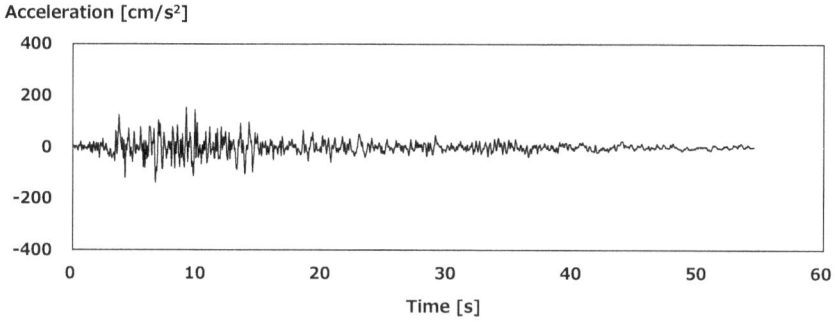

Figure 11.2 Chart indicating relation between acceleration and time: Taft EW (1952).

Acceleration [cm/s²]

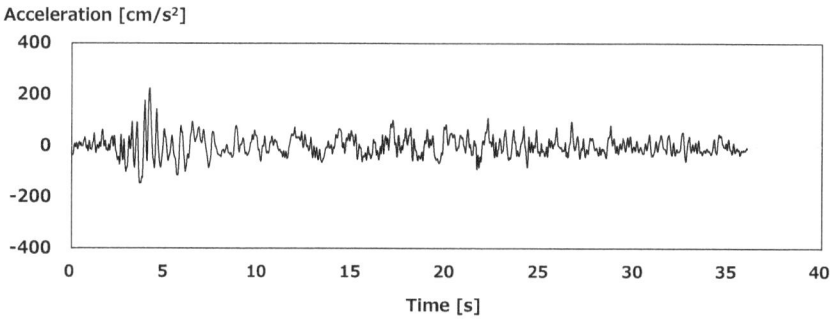

Figure 11.3 Chart indicating relation between acceleration and time: Hachinohe NS (1968).

Acceleration [cm/s²]

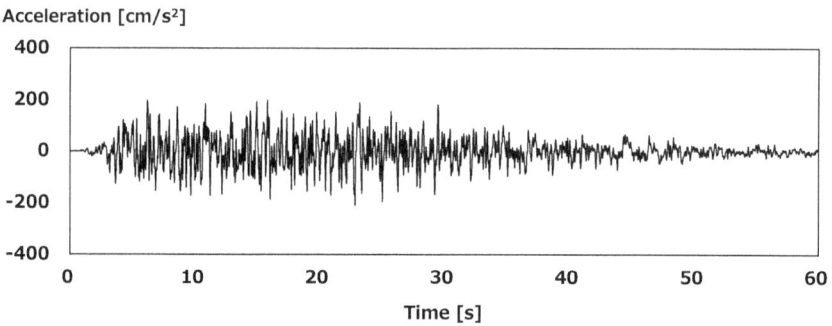

Figure 11.4 Chart indicating relation between acceleration and time: BCJ-L1.

model where six spring oil dampers (modeled using the improved model) were installed in six different locations (this analysis model is abbreviated "BODV"). In addition, as indicated in Figure 11.6, each of these reinforcement components was installed to the point mass of the structural model

Acceleration [cm/s²]

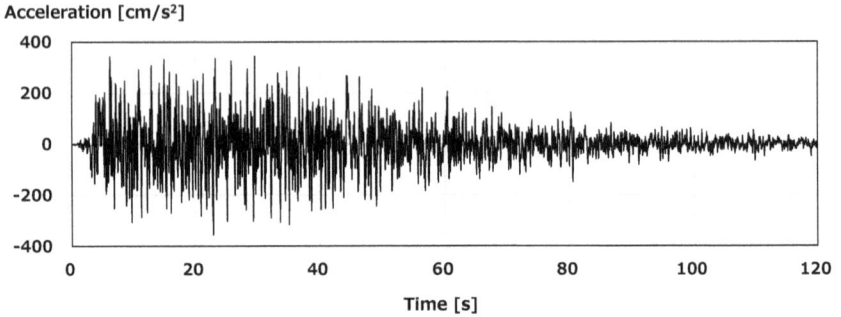

Figure 11.5 Chart indicating relation between acceleration and time: BCJ-L2.

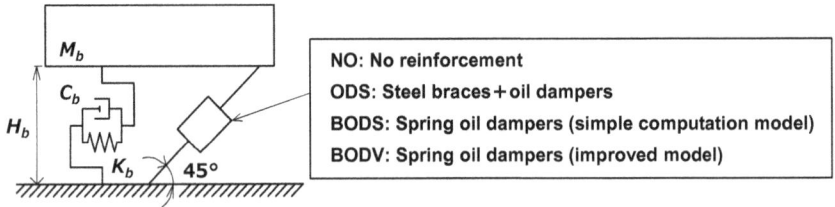

NO: No reinforcement

ODS: Steel braces + oil dampers

BODS: Spring oil dampers (simple computation model)

BODV: Spring oil dampers (improved model)

Figure 11.6 Analysis cases specific to analysis models and reinforcement types.

at a 45° angle. Modeling of the spring oil dampers was done by combining the Maxwell model, linear spring model, Voigt model, and friction damper model, and the specifications of each of these models are provided in Table 11.3. For this analysis, the viscoelastic springs' dependence on displacement was not taken into consideration. For the steel brace design, it was modeled as truss elements while only considering their axial deformation, having the dimensions of L-90×90×6 and the stiffness and yielding load similar to the

Table 11.3 Spring oil damper specifications

Spring oil damper	Maximum load [kN]		750
	Oil damper	Maximum load [kN]	500
		Maximum velocity [mm/s]	500
		Relief load [kN]	400
		Relief velocity [mm/s]	32
		Primary attenuation coefficient [kNs/mm]	12.5
		Secondary attenuation coefficient [kNs/mm]	0.21
		Stiffness [kN/mm]	140
	Spring unit	Stiffness [kN/mm]	25
		Attenuation coefficient [kNs/mm]	0.70
		Friction force [kN]	250

stiffness and friction force of the spring units. In terms of their restoration force characteristics, as their installation length would be long, with a slenderness ratio of about 300, the bilinear-slip model was used.[1, 2] As for their post-yield stiffness under load in the bilinear-slip model, it was assumed to be the same as the initial stiffness. The model design would also incorporate a steel tube being attached to each oil damper and spring oil damper in ODS, BODS, and BODV for length adjustment, each of which having an outer diameter of 267.4 mm and a thickness of 12.7 mm. So, to mimic these steel tubes in the actual simulation, springs with an axial stiffness of 272.08 [kN/mm] were serially connected in each of the aforementioned models. Table 11.4 indicates the results of the complex eigenvalue analysis performed in each of the analysis cases.

Based on the results provided in Table 11.4, it was determined that each of the reinforced models had exhibited a shorter natural period than the non-reinforced model, along with a higher attenuation constant. The data also shows that BODS and BODV had a longer natural period than ODS. This is attributable to the fact that the amount of stiffness support added to the single degree-of-freedom model by each of BODS and BODV was less than that added by ODS, as the springs mimicking the characteristics of the steel tubes were serially connected to the damper models instead of the actual steel tubes. The equivalent viscous attenuation coefficient of the Voigt model used in the improved model was calculated based on the primary natural period data, which was obtained from the formula shown in Figure 10.13 of Section 10.3.2 and the complex eigenvalue analysis results related to BODS shown in Table 11.4.

11.3 ANALYSIS RESULTS

The results of this time history response analysis performed with the single degree-of-freedom model are shown in Table 11.5 in terms of maximum relative story drift angles specific to the analysis cases and in Table 11.6 in terms of maximum absolute response acceleration. In addition, Table 11.7 provides the breakdown of the energy absorption rates measured following the BCJ-L2 wave input, while Figure 11.7 presents graphs indicating the relation between story shear force and relative story drift angle, and Figure 11.8

Table 11.4 Complex eigenvalue analysis results

Analysis case	Natural period [s]	Attenuation constant [%]
NO	1.176	0.02
ODS	0.347	0.147
BODS	0.406	0.195
BODV	0.399	0.213

Table 11.5 Maximum relative story drift angle data from different analysis cases [rad]

		Analysis cases			
Lv.	Input waveform	NO	ODS	BODS	BODV
1	El Centro	1/63	1/279	1/213	1/254
	Taft	1/72	1/272	1/228	1/263
	Hachinohe	1/78	1/559	1/527	1/565
	BCJ-L1	1/52	1/287	1/255	1/310
2	El Centro	1/35	1/82	1/92	1/98
	Taft	1/46	1/85	1/84	1/89
	Hachinohe	1/33	1/205	1/211	1/221
	BCJ-L2	1/17	1/98	1/107	1/118

Table 11.6 Maximum absolute response acceleration data from different analysis cases [cm/s^2]

		Analysis cases			
Lv.	Input waveforms	NO	ODS	BODS	BODV
1	El Centro	239.17	279.19	287.97	275.51
	Taft	225.77	283.28	281.91	271.86
	Hachinohe	212.81	217.13	204.07	197.65
	BCJ-L1	240.91	273.97	272.73	259.89
2	El Centro	248.75	279.19	395.22	387.20
	Taft	244.36	411.20	408.27	401.56
	Hachinohe	255.76	299.84	288.01	284.70
	BCJ-L2	261.11	390.69	368.27	354.89

Table 11.7 Breakdown of energy absorption rate data: BCJ-L2 [%]

	Analysis cases			
Type	NO	ODS	BODS	BODV
E_d/E_i	26.64	3.94	6.88	7.69
E_m/E_i	–	91.19	83.56	74.0
E_s/E_i	72.89	4.05	0.005	0.004
E_v/E_i	–	–	–	11.21
E_h/E_i	–	–	9.55	7.19

where: E_i = energy input; E_d = energy absorption due to structural attenuation; E_m = energy absorption according to the Maxwell model; E_s = energy absorption according to the spring model; E_v = energy absorption according to the Voigt model; and E_h = energy absorption according to the friction damper model.

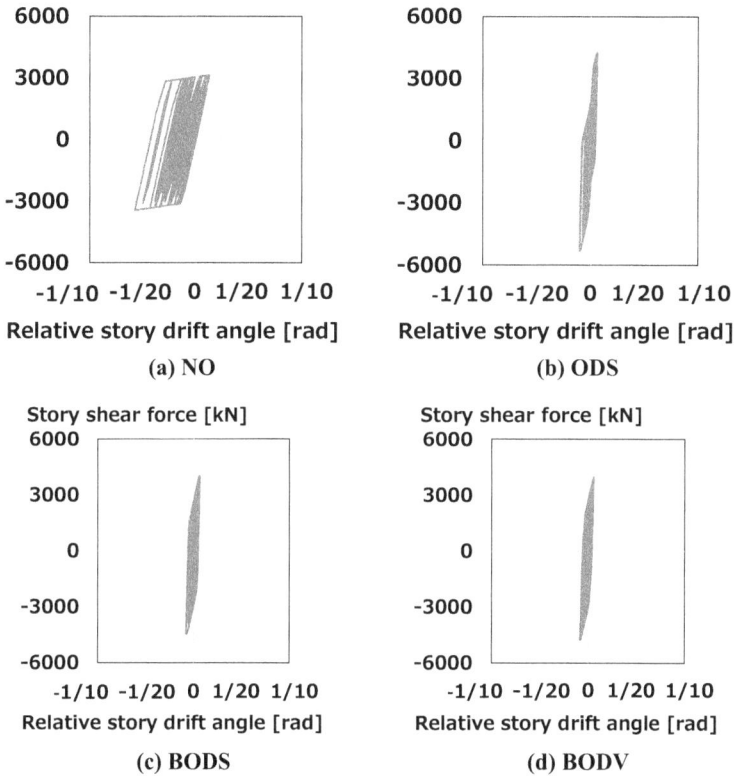

Figure 11.7 Chart indicating relation between story shear force and relative story drift angle: BCJ-L2.

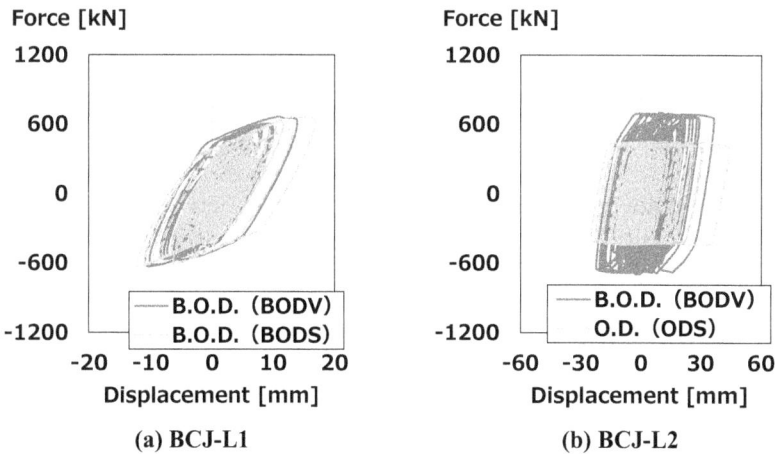

Figure 11.8 Chart indicating relation between force and displacement related to damper components.

presents graphs indicating the force-displacement relation observed with the damper components, including the steel tubes for length adjustment, in response to the BCJ-L1 and BCJ-L2 wave input. Further in Figure 11.8, the chart entitled "BCJ-L1" is a superimposed graph indicating the force-displacement relation observed with the spring oil dampers in BODS and BODV, while the chart entitled "BCJ-L2" is a superimposed graph indicating the force-displacement relation observed with the oil dampers in ODS and the spring oil dampers in BODV.

Comparative before-after analysis of the maximum relative story drift angle data shown in Table 11.5 revealed that the maximum relative story drift angles were significantly reduced by the reinforcement across all analysis cases. Upon side-by-side examination of the post-reinforcement cases, it was determined that the maximum relative story drift angles were more or less the same across all cases. However, the data that resulted from the BCJ-L2 wave input – which was less prone to the effects of frequency components and had the tendency to increase on the displacement response spectrum along with the natural period – demonstrated that the relative story drift angles were smaller in BODS and BODV, in which spring oil dampers had been installed, compared to ODS. Furthermore, examination of the Lv. 1 and Lv. 2 analysis cases revealed that many of them had smaller response values in BODV than they did in ODS. However, BODS apparently had larger relative story drift angles than did ODS across all Lv. 1 analysis cases. When BODS and BODV were compared, BODV – which used the improved model that took into account the energy absorption capacity of the viscoelastic springs based on the Voigt model – apparently had smaller relative story drift angles than did BODS across all analysis cases, especially in response to Lv. 1-equivalent acceleration waveform input. In addition, the before-after comparison data provided in Table 11.6 in terms of maximum absolute response acceleration demonstrated that the response acceleration had increased following the reinforcement and that the response acceleration was roughly the same across all post-reinforcement analysis cases of ODS, BODS, and BODV. Moreover, a side-by-side examination of BODS and BODV in terms of response acceleration revealed that BODV had yielded slightly lower figures than did BODS across all analysis cases. As for the breakdown of the energy absorption rates shown in Table 11.7, the analysis cases of BODS and BODV exhibited lower energy absorption rates in the Maxwell model compared to ODS. In terms of Figure 11.7, a comparative analysis of the graphs indicating the relation between story shear force and relative story drift angle in ODS, BODS, and BODV revealed that the story shear force and the energy absorption in ODS in the second and fourth quadrants were smaller than in BODS and BODV, due to the characteristics of the restoration force unique to the brace components simulated using the bilinear-slip model. As for Figure 11.8, the graphs indicating the force-displacement relation observed with the spring oil dampers in response to the BCJ-L1 input demonstrated that BODV – which simulated the spring oil dampers using the improved model that factored in the attenuation characteristics of the viscoelastic springs incorporated therein – had exhibited greater force generated due to their viscous property than would BODS. Further, examination of the superimposed BCJ-L2 graph indicating the force-displacement relation of data specific to the oil dampers and the spring oil dampers demonstrated that the spring oil dampers had generated more force and absorbed more energy than did the conventional oil dampers. In terms of comparison between the simple computation model and the improved model in simulating the viscoelastic springs integrated into the spring oil dampers, no significant difference was observed in response values following the Lv. 2-equivalent acceleration waveform input. In addition, as the analysis protocol assumed that each of the reinforcement components would be installed in six different locations, ODS had a total of 12 reinforcement points, while BODS and BODV each had six, indicating that the use of spring oil dampers could reduce the number of reinforcement points while enhancing the structural performance post-reinforcement.

where: E_i = energy input; E_d = energy absorption due to structural attenuation; E_m = energy absorption according to the Maxwell model; E_s = energy absorption according to the spring model; E_v = energy absorption according to the Voigt model; and E_h = energy absorption according to the friction damper model.

11.4 SUMMARY

In this chapter, the single degree-of-freedom model was built simulating an actual factory building to perform time history response analysis, involving the two types of spring oil dampers – one with the attenuation characteristics of the built-in viscoelastic spring and the other without – represented in the different analysis models to determine how the response values would be impacted by the different models and to verify the efficacy of the spring oil dampers.

First, Section 11.2 provided the specifications of all the analysis models along with the analysis conditions.

Then, Section 11.3 explained the results of the analysis performed using the four different models, namely the model without any reinforcement (NO), the model reinforced by oil dampers and steel braces (ODS), the model reinforced by spring oil dampers and simulated using the simple computation model (BODS), and the model reinforced by spring oil dampers and simulated using the improved model (BODV). Examination of the viscoelastic springs integrated into the spring oil dampers, simulated using both the simple computation model and the improved model, did not reveal any significant difference in response values following the Lv. 2-equivalent acceleration waveform input. In addition, as the analysis protocol assumed that each of the reinforcement components would be installed in six different locations across the structure, ODS had a total of 12 reinforcement points, while BODS and BODV each had six, demonstrating that the use of spring oil dampers could reduce the number of reinforcement points while improving the structural performance post-reinforcement.

REFERENCES

1. M Nakahara, T Shigenobu, A Watanabe: "Seismic response and restoration force characteristics of bracing structures, part 1: Equivalent elasto-plastic-slip system", *Summaries of Scientific Lectures from the Architecture Institute of Japan (AIJ) Annual Meetings, C, Structure II*, pp.1037–1038, July 1988. (in Japanese)
2. A Watanabe, T Shigenobu, M Nakahara: "Seismic response and restoration force characteristics of bracing structures, part 2: Distributed-element-type restoration force model", *Summaries of Scientific Lectures from the Architecture Institute of Japan (AIJ) Annual Meetings, C, Structure II*, pp.1039–1040, July 1988. (in Japanese)

Chapter 12

Converted I_s and q values of building incorporating stiffness-supported oil dampers

12.1 INTRODUCTION

This chapter describes a simple method that can be employed to assess the efficacy of spring oil dampers in response reduction through seismic attenuation, involving static assessment techniques based on the seismic index of structure (I_s) and ultimate horizontal resistant force (q) as specified in the seismic diagnosis. First, Section 12.2 provides the specifications of the building model used in the assessment. Then, Section 12.3 explains how converted I_s and q values were calculated, reflecting the response reduction resulting from the spring oil dampers' seismic attenuation. Finally, Section 12.4 demonstrates the process of calculating the converted I_s and q values relating to a frame model that simulated the building model as specified in Section 12.2, incorporating spring oil dampers for reinforcement, and compares them to the results of the time history response analysis previously performed to determine the validity of the converted I_s and q values calculated using the proposed method.

12.2 BUILDING MODEL SPECIFICATIONS AND SEISMIC PERFORMANCE

Based on the structural model of the steel-framed gymnasium as specified by the Japan Building Disaster Prevention Association (JBDPA),[1] a building model was created that would be used as a warehouse. This model is of a single-story building that has cross bracing across its longitudinal frames and a pitched rigid-frame structure widthwise. It has a span of 16 m widthwise and 10 spans each measuring 5.8 m lengthwise, and an eave height of 6.5 m and a building height of 9.3 m. Figure 12.1 is the building model's roof plan while Figures 12.2 and 12.3 indicate its framing elevation.

Table 12.1 is a list of the components that comprise the building model. As indicated in Table 12.2, SS400 was selected as the material of which to construct the model. Table 12.3 provides a list of the dead loads that form part of the building model. In this connection, the finish weight of

DOI: 10.1201/9781003290261-14

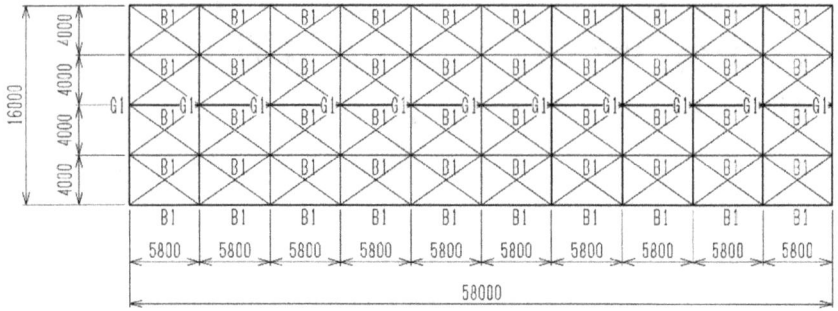

Figure 12.1 Roof plan.

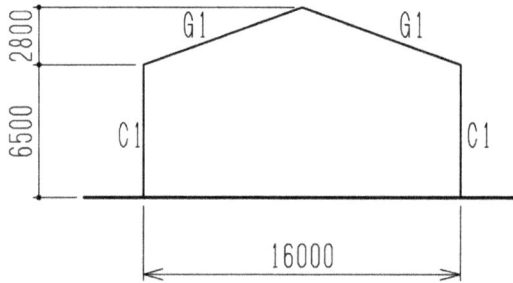

Figure 12.2 Widthwise framing elevation.

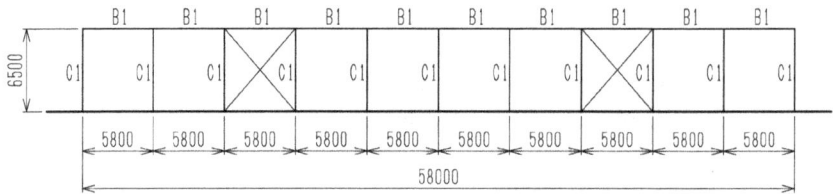

Figure 12.3 Lengthwise framing elevation.

Table 12.1 List of components

Component types and symbols	Structural steel
Column C1	H—340×250×9×14
Girder (lengthwise) G1	H—340×250×9×14
Beam (widthwise) B1	H—300×150×6.5×9
Brace	L—60×60×5

Table 12.2 Material type and strength

Building material	Material strength F [N/mm²]
SS400	235

Table 12.3 List of dead loads

Types	Components	Weight per component [kN/m²]	Total weight [kN/m²]
Roof	Corrugated steel plates	0.075	0.64 [kN/m²]
	Wood wool board substrate	0.145	
	Purlins	0.07	
	Steel frames	0.30	
	Other	0.05	
Gable-side top-layer walls	Gable-side finish	0.490	0.685 [kN/m²]
	Gable-side substrate	0.195	
Top-layer walls			1.0 [kN/m²]

the column and beam components was set to 0.1 kN per unit area. For the tasks of model creation and weight calculation, SNAP LE Ver.7 was used. As for the weight of the roof, the figures indicated in Table 12.3 were applied as surface loads, while the weight of the walls was dealt with by multiplying the figures from Table 12.3 by the wall area to obtain their weight, then dividing a half of the weight by the beam length, and applying it as uniformly distributed loads to the beam components. An image of the frame model constructed by the aforementioned process is provided in Figure 12.4. The mass of the created analysis model was 159.11 t. As the main objective of this chapter is to explain the method of calculating the converted I_s and q values related to the building that is reinforced using

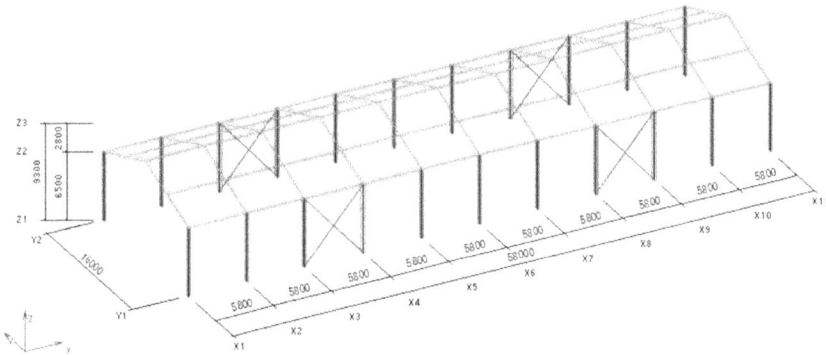

Figure 12.4 Frame model drawing: Standard model.

the spring oil dampers, the analysis focuses only on the model's behavior lengthwise. In terms of the bracing components, since they are made of angle steel with a high slenderness ratio, the analysis only considers their tensile strength. The braces were assembled so that the resistant force of their joints would be greater than their absolute resistant force.

Load increase analysis was performed on the frame model to calculate its ultimate horizontal resistant force. SNAP-LE Ver.7 was the software used for this analysis task. For this analysis model, it was assumed that the column bases had pin joint structures, the connecting points had rigid joints, and the joining panel zones were also rigid. In addition, the roof surfaces were assumed to have a rigid floor, while their torsion caused by the horizontal force would be ignored. In terms of the flexural capacity of the column components, the MN model was applied as it would reflect the change in flexural capacity caused by the axial force. Shear strength was calculated based on a uniaxial spring model. The beam components were also represented using a uniaxial spring model to calculate their flexural strength and shear strength. The bracing components were modeled as trusses only considering their axial deformation. In terms of proof stress, the compressive strength was assumed to be 0, while the tensile strength was calculated by multiplying the cross-sectional area of the bracing components by the material strength. As for the load increase analysis protocol, the analysis was performed until the relative story drift angle reached 1/20 [rad], and the effect of P-Δ was considered by first determining the overturning moment that gravity would cause to the story, which was calculated based on the story mass and the relative story drift angle, and applying a horizontal force equivalent to the moment to the story's center of mass.[2, 3] Figure 12.5 provides the results of this load increase analysis in the form of a graph indicating the relation between force and displacement. The yield point shown in Figure 12.5 was defined as the step where all the bracing components had yielded, while the safety limit was defined as the step where plastic hinges had occurred to all the beam components on both ends in the X direction. Figures 12.6 and 12.7 present drawings of ductility factor distribution across the structure at the yield point and the safety limit, respectively. Based on the force-displacement curve in Figure 12.5, the standard model design criteria at the safety limit included a relative story drift angle of 1/30 rad. The I_s and q values of the standard model were then determined as explained below. As for the ultimate horizontal resistant force that would be applied to calculate I_s and q, the force at the yield point where the bracing components would yield was used as described by JDBPA.[1] In terms of toughness index F, since the angle steel frames – whose joints' resistant force would be greater than their absolute resistant force – would bear the horizontal force, F was set to 3.3. The coefficient F_{es} – a product of the modulus of rigidity and the eccentricity ratio – and the seismic zoning factor Z were each set to 1. For the assessment of the standard model's seismic performance, as the above parameter setting led to I_s=0.90 and q=1.09, it

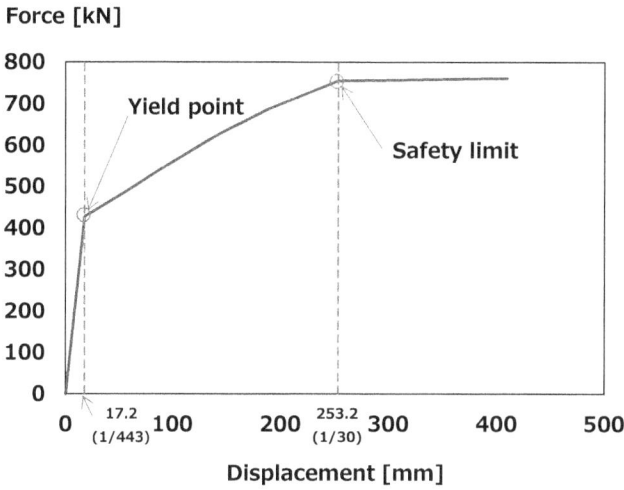

Figure 12.5 Chart indicating relation between force and displacement related to frame model: Standard model.

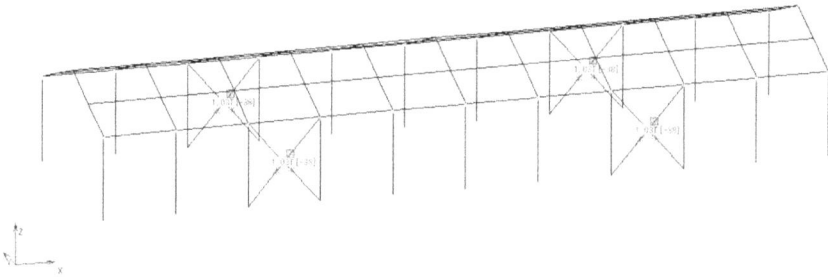

Figure 12.6 Ductility factor distribution diagram: Yield point.

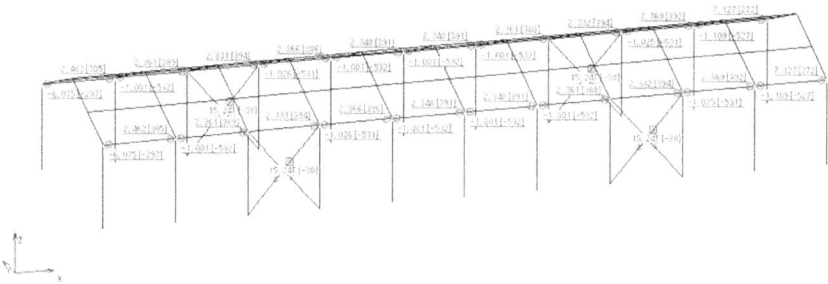

Figure 12.7 Ductility factor distribution diagram: Safety limit.

was determined that the risk of the model structure collapsing or being destroyed in response to seismic vibration and shock would be low (assuming $I_s \geq 0.6$ and $q \geq 1.0$).[1]

- Each red square shown in Figure 12.6 denotes a plastic hinge that is caused by the tensile force acting on the bracing components. Each red asterisk denotes a plastic hinge caused by the compressive force acting on the bracing components, while each numeric value denotes the ductility factor.
- Each red circle shown in Figure 12.7 denotes a plastic hinge that is caused by the bending force acting on the beam components, while each numeric value denotes the ductility factor [number of the step where the yield occurred]. The meaning of the red squares and asterisks is the same as that explained for Figure 12.6.

$$E_0 = \frac{Q_u F}{WA} = \frac{425.74 \times 3.3}{1560.34 \times 1.0} = 0.90$$

$$I_s = \frac{E_0}{F_{es} Z R_t} = \frac{0.90}{1.0 \times 1.0 \times 1.0} = 0.90$$

$$q = \frac{Q_u}{0.25 F_{es} WAZR_t} = \frac{425.74}{0.25 \times 1.0 \times 1560.34 \times 1.0 \times 1.0 \times 1.0} = 1.09$$

where: E_0 = seismic performance index; Q_u = ultimate horizontal resistant force [kN]; F = toughness index calculated based on the plastic deformation performance of components and joints specific to each story and direction; W = building mass [kN]; A = heightwise distribution of story shear force as defined in the Order for Enforcement of the Building Standards Act; I_s = seismic index of structure; F_{es} = coefficient that is a product of the modulus of rigidity and the eccentricity ratio, expressed as $F_{es} = F_s F_e$; F_s = coefficient determined by the modulus of rigidity calculated from the relative story drift angle; F_e = coefficient determined by the eccentricity ratio when the planar asymmetry between proof stress and mass distribution is significant; Z = seismic zoning factor as defined in the Order for Enforcement of the Building Standards Act; R_t = vibration characteristic factor as defined in the Order for Enforcement of the Building Standards Act; and q = ultimate horizontal resistant force index.

Then, time history response analysis was performed by applying different acceleration waveforms to the frame model that had been constructed as shown in Figure 12.4. The software used for this analysis was SNAP-LE Ver.7. In terms of computation method for this time history response analysis, the Newmark-β method (β=0.25) was applied, while the structural attenuation of the building's main structure was assumed to manifest as momentary stiffness proportional damping (h=2%). For this analysis model, it was assumed that the column bases had pin joint structures, the connecting points had rigid joints, and the joining panel zones were also rigid. In addition, the roof surfaces were assumed to have a rigid floor, while their torsion caused by the horizontal force would be ignored. Due to the foregoing assumption, the story height of the roof surfaces turned out to be 7620 mm. In terms of the flexural capacity of the column components, the MN model was applied as it would reflect the change in flexural capacity caused by the axial force. Shear strength was calculated based on a uniaxial spring model. As for the restoration force characteristics of the column components, the MN model was applied for bending while the bilinear-slip model was applied for shearing. The beam components were represented using a uniaxial model to calculate their flexural strength and shear strength. In terms of the restoration force characteristics of the beam components, the stiffness-reduced model A (normal bilinear) was applied for bending, while the bilinear-slip model was applied for shearing. As for the proof stress of the bracing components, the compressive strength was assumed to be 0, while the tensile strength was calculated by multiplying the cross-sectional area of the bracing components by the material strength. The restoration force characteristics of the bracing components in response to axial force were expressed as a bilinear-slip model. The post-yield stiffness under load was assumed to be the same as the initial stiffness. For this time history response analysis, a total of five different acceleration waveforms were applied. The details of these waveforms such as their names, maximum acceleration, and maximum velocity are provided in Table 12.4. As the acceleration waveforms that were applied here are identical to those provided in Figures 11.1 through 11.3 and Figure 11.5, they are restated. For each observed wave data besides BCJ-L2, a maximum velocity of 50 cm/s was uniformly applied, while the original waveform model of BCJ-L2 was used in the Lv. 2 analysis case to transmit the acceleration waveform to the structure's foundation accordingly. As this analysis mainly focused on determining the model's response values during major earthquakes, only the Lv. 2 acceleration waveforms were applied.

Table 12.4 List of input waveforms

Type	Name	Maximum acceleration [cm/s²]	Maximum velocity [cm/s]
Actual observed waves	El Centro NS (1940)	341.7	33.66
	Taft EW (1952)	175.9	17.69
	Hachinohe NS (1968)	225.0	31.44
Simulated seismic waves	BCJ-L2 (Artificial)	355.7	57.40

Table 12.5 Maximum relative story drift angle data
from different analysis cases: Lv. 2 [rad]

El Centro	Taft	Hachinohe	BCJ-L2
1/51	1/61	1/98	1/70

Table 12.6 Maximum absolute response acceleration data
from different analysis cases: Lv. 2 [cm/s²]

El Centro	Taft	Hachinohe	BCJ-L2
398.29	375.33	328.53	360.54

The results of the time history response analysis are shown in Table 12.5 as maximum relative story drift angles specific to the input acceleration waveforms, and in Table 12.6 as maximum absolute response acceleration. Figure 12.8 presents the ductility factor distribution diagram in response to the El-Cento NS (1940) input, while Figure 12.9 indicates the force-displacement graph of a sample of the bracing components, and Figure 12.10 provides a graph indicating the relation between story shear force and relative story drift angle specific to each story. Examination of Table 12.5 revealed the maximum relative story drift angle was only 1/51 of the figure that arose based on the El-Cento NS (1940) input. As the relative story drift angle at the yield point was 1/443 during the load increase analysis, the story-specific ductility factor turned out to be 8.69 which was rather significant. However, this was because the bracing's rigidity that would bear the horizontal force accounted for most of the stiffness of the entire standard model, resulting in the standard model's deformation to a more severe degree after the bracing had yielded. Also, analysis of the ductility factor distribution diagram in Figure 12.8 indicated that several of the beam and bracing components had yielded. In this connection, the maximum ductility factor was roughly 1.5 for the beam components and 8.9 for the bracing components. As for the standard model's seismic performance,

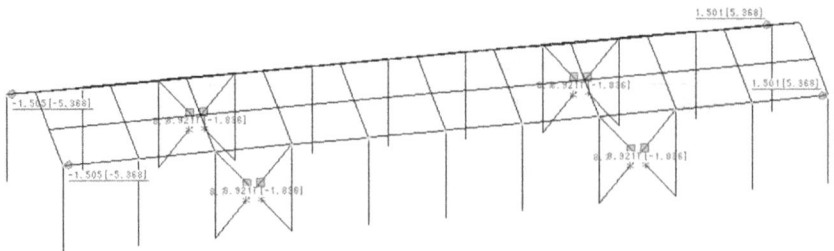

Figure 12.8 Ductility factor distribution diagram: El Centro.

Force [kN]

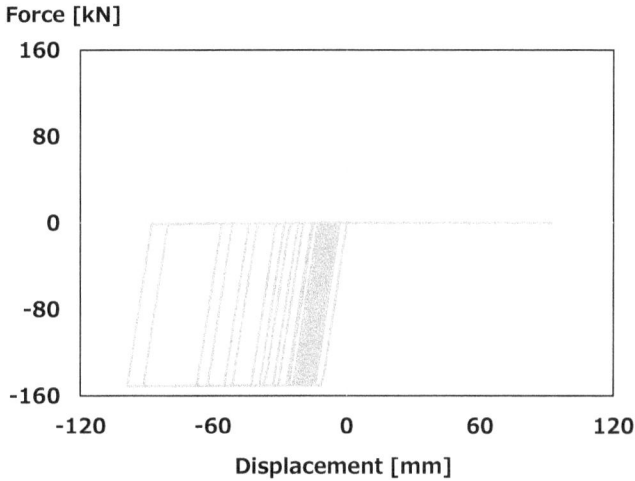

Figure 12.9 Chart indicating relation between force and displacement related to bracing components: El Centro.

Story shear force [kN]

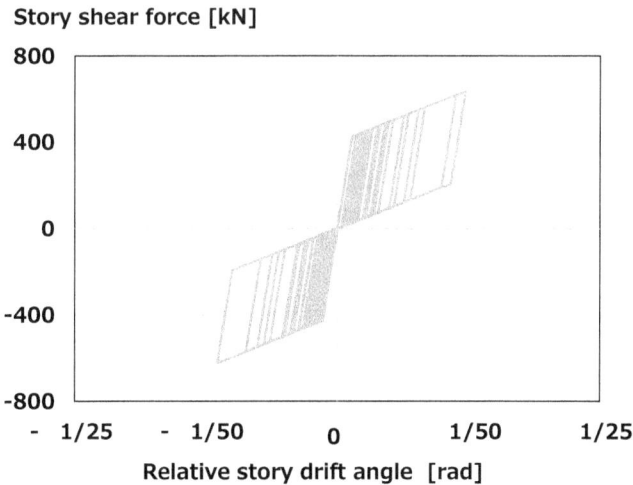

Figure 12.10 Chart indicating relation between story-specific story shear force and relative story drift angle: El Centro.

the risk of it collapsing or being destroyed in response to seismic vibration and shock would be low, since the maximum relative story drift angle that resulted from the Lv. 2 acceleration waveform input did not exceed 1/30 of the design criterion at the safety limit. In addition, examination of the force-displacement graph of the bracing components in Figure 12.9 indicated that the shape of their hysteretic performance manifested as a bilinear-slip model that would only bear the load in the tensile direction. Also, analysis of the graph indicating the relation between story shear force and relative story drift angle specific to each story in Figure 12.10 revealed that the hysteretic data took the shape of the bracing components' performance expressed as a bilinear-slip model, with a slight tilt proportional to the additional stiffness of the rigid-frame structure.

12.3 SIMPLE COMPUTATION METHOD FOR FINDING CONVERTED I_s AND Q VALUES

In this section, a computation method is proposed that can be used to calculate in a simplified manner the converted I_s and q values that reflect the seismic attenuation effect of a structure, by first calculating D_h – a coefficient having a significance similar to the attenuation correction factor in reference to the response spectrum – and multiplying the reciprocal of the calculated reduction coefficient by the seismic index of structure (I_s) and ultimate horizontal resistant force (q), respectively, based on the assumption that an acceleration that arises in a structure is subsequently reduced by attenuation in the calculation of response and limit strength. For the purpose of this dissertation, D_h is referred to as the response reduction factor. The process of calculating the converted I_s and q values is explained in the flowchart below. Step-by-step instructions on how to calculate the response reduction factor D_h based on Figure 12.11 and flowchart are as follows:

(1) Calculate the equivalent stiffness of the frame model at the safety limit pre- and post-reinforcement based on the load increase analysis to determine the additional stiffness provided by the spring oil dampers.
(2) Assuming that an additional attenuation coefficient will be added in proportion to the added stiffness, calculate the attenuation constant from the additional attenuation coefficient using the formula (12.1).[4]
(3) Plug the attenuation constant into the formula (12.2)[5] to obtain the response reduction factor D_h.

As shown in the formulas (12.3) and (12.4), the attenuation coefficient of the spring oil dampers is defined as the sum of the attenuation coefficients of the oil dampers and the viscoelastic springs, so the variation in attenuation coefficient caused by the stiffness of the oil dampers and the steel tubes for length adjustment and also by vibration frequency is considered by this method. Also, as the attenuation characteristics of the oil dampers are bilinear, the attenuation coefficient c_{od} is converted to an equivalent linear value in this calculation as shown in Figure 12.12.[5]

$$h_{bod} = \frac{c_{bod}}{2\sqrt{mk}} \qquad (12.1)$$

where: h_{bod} = attenuation constant added to the frame model by the spring oil dampers; c_{bod} = spring oil dampers' attenuation coefficient [kN•s/m], m = frame model's mass [t], and k = frame model's stiffness after reinforcement [kN/m].

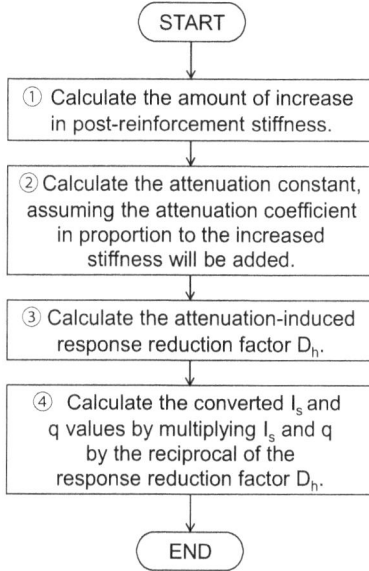

```
              ┌─────────────┐
              │    START    │
              └─────────────┘
                     │
                     ▼
     ┌───────────────────────────────┐
     │ ① Calculate the amount of      │
     │ increase in post-reinforcement │
     │ stiffness.                     │
     └───────────────────────────────┘
                     │
                     ▼
     ┌───────────────────────────────┐
     │ ② Calculate the attenuation    │
     │ constant, assuming the         │
     │ attenuation coefficient        │
     │ in proportion to the increased │
     │ stiffness will be added.       │
     └───────────────────────────────┘
                     │
                     ▼
     ┌───────────────────────────────┐
     │ ③ Calculate the attenuation-   │
     │ induced response reduction     │
     │ factor Dₕ.                     │
     └───────────────────────────────┘
                     │
                     ▼
     ┌───────────────────────────────┐
     │ ④ Calculate the converted Iₛ   │
     │ and q values by multiplying Iₛ │
     │ and q by the reciprocal of the │
     │ response reduction factor Dₕ.  │
     └───────────────────────────────┘
                     │
                     ▼
              ┌─────────────┐
              │    END      │
              └─────────────┘
```

Figure 12.11 Flowchart for calculating converted I_s and q values.

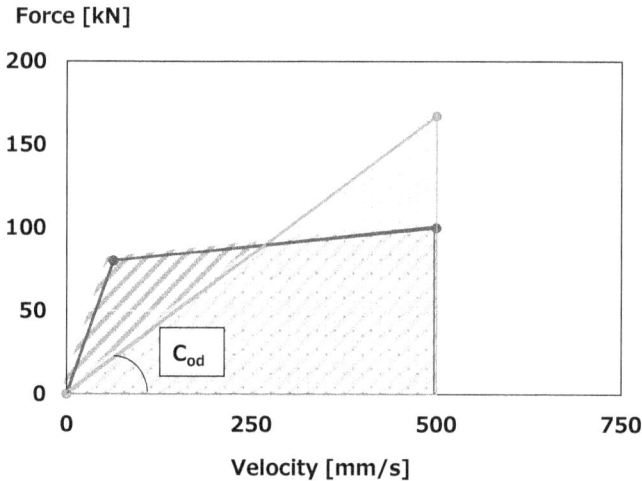

Figure 12.12 Attenuation characteristic substitution.

$$D_h = \sqrt{\frac{1+25h_0}{1+25h_{eq}}} \tag{12.2}$$

where: h_0 = initial attenuation; and h_{eq} = equivalent attenuation.

$$c_{bod} = \frac{c_{od'} + c_b}{1 + \left(\{c_{od'} + c_b\}\omega / k_p\right)^2}\left(\frac{k_{eqr} - k_{eq}}{k_{eqr}}\right) \tag{12.3}$$

where: $c_{od'}$ = oil dampers' attenuation coefficient [kN•s/m] reflecting the effects of stiffness and vibration frequency; c_b = viscoelastic springs' attenuation coefficient [kN•s/m], ω = circular frequency ($= 2\pi f$) [rad/s]; k_p = length-adjustment steel tube's rigidity [kN/m]; k_{eqr} = frame model's equivalent stiffness [kN/m] at the safety limit after reinforcement by spring oil dampers; and k_{eq} = frame model's equivalent stiffness [kN/m] at the safety limit before reinforcement.

$$c_{od'} = \frac{c_{od}}{1 + \left(c_{od}\omega / k_{od}\right)^2} \tag{12.4}$$

where: c_{od} = oil dampers' attenuation coefficient [kN•s/m]; and k_{od} = oil dampers' stiffness [kN/m].

$$I_{sd} = \frac{1}{Dh}I_s \tag{12.5}$$

where: I_{sd} = converted I_s.

$$q_d = \frac{1}{Dh}q \tag{12.6}$$

where: q_d = converted q.

12.4 COMPUTATION OF CONVERTED I_s AND Q VALUES AND VERIFICATION OF THEIR VALIDITY

The standard model that was introduced in Section 12.2 might only pose a low risk of collapsing or being destroyed in response to seismic vibration or shock, as its I_s was above 0.6 and its q was above 1.0. Therefore, as the next step in this progression, another model with reduced bracing was simulated, in which the number of locations where the braces were installed would be reduced from four, as shown in the standard model in Figure 12.4, to two, with reinforcement by spring oil dampers, to calculate its converted I_s and q values. Figure 12.13 provides a diagram of this bracing-reduced version of the frame model. For the bracing-reduced model, I_s and q were calculated as explained below. For this calculation, the ultimate horizontal

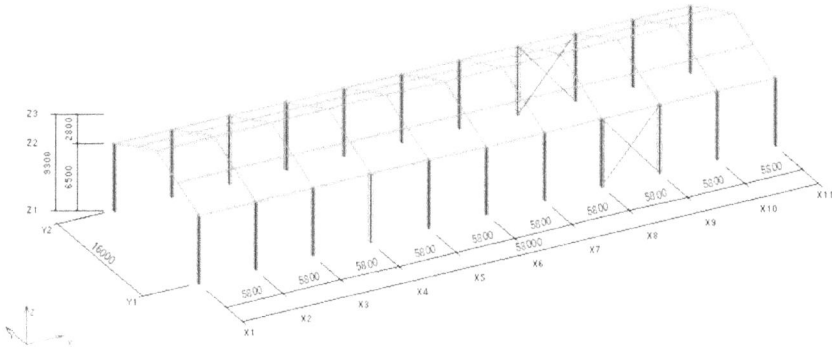

Figure 12.13 Frame model drawing: Bracing-reduced model.

resistant force was computed based on the load increase analysis, and the same analysis conditions were applied, so they are not restated here. The bracing-reduced model's I_s was 0.47 while its q was 0.57. Hence, in terms of the assessment of this bracing-reduced model's seismic performance based on the computed values, there would be a significant risk that the structure might collapse or be destroyed in response to seismic vibration and shock (assuming $0.3 \leq I_s < 0.6$ and $0.5 \leq q < 1.0$).[1]

$$E_0 = \frac{Q_u F}{WA} = \frac{222.98 \times 3.3}{1560.34 \times 1.0} = 0.47$$

$$I_s = \frac{E_0}{F_{es} Z R_t} = \frac{0.47}{1.0 \times 1.0 \times 1.0} = 0.47$$

$$q = \frac{Q_u}{0.25 F_{es} WAZR_t} = \frac{222.98}{0.25 \times 1.0 \times 1560.34 \times 1.0 \times 1.0 \times 1.0} = 0.57$$

It was assumed that the spring oil dampers having the specifications as set forth in Table 12.8 would be installed in the frame model for reinforcement. A diagram of the spring-reinforced frame model is presented in Figure 12.14, and the results of the model's eigenvalue analysis are shown in Table 12.7. The model of the spring oil dampers used for this eigenvalue analysis was a simplified version that would not consider the attenuation effect of the viscoelastic springs incorporated into the dampers. Table 12.7 also provides the analysis results relating to the standard model and the bracing-reduced model for comparison. In terms of the attenuation specific to each of the model's components that was considered in the eigenvalue analysis, the value was set to 2% of what would be used in the initial-stiffness-proportionate model for each component. The spring oil damper specifications are provided in Table 12.8. The spring oil dampers' maximum load and stiffness were adjusted so that they would approximate the yield load and rigidity of

Table 12.7 Natural periods and attenuation constants: Complex eigenvalue analysis

Analysis cases	Natural periods [s]	Attenuation constants [%]
Spring-reinforced model	0.387	0.17
Standard model	0.360	0.02
Bracing-reduced model	0.500	0.02

Table 12.8 Spring oil damper specifications

Spring oil damper	Maximum load [kN]		150
	Oil damper	Maximum load [kN]	100
		Maximum velocity [mm/s]	500
		Relief load [kN]	80
		Relief velocity [mm/s]	64
		Primary attenuation coefficient [kNs/mm]	12.5
		Secondary attenuation coefficient [kNs/mm]	0.46
		Stiffness [kN/mm]	55
	Spring unit	Stiffness [kN/mm]	10
		Attenuation coefficient [kNs/mm]	0.35
		Friction force [kN]	50

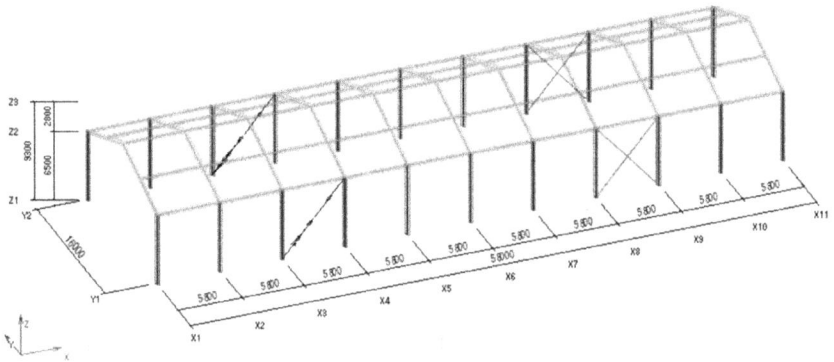

Figure 12.14 Frame model drawing: Spring-reinforced model.

the bracing components. As for the attenuation coefficient of the viscoelastic springs, the value was calculated based on the eigenvalue analysis results indicated in Table 12.7 using the formula as specified in Figure 10.13. While this chapter mentions the response value resulting from the Lv. 2 acceleration input, an improved version of the spring oil damper model was used to calculate the response reduction factor D_h that would reflect the viscoelastic springs' attenuation effect. Figure 12.15 provides a force-displacement graph obtained during the load increase analysis related to the bracing-reduced

Force [kN]

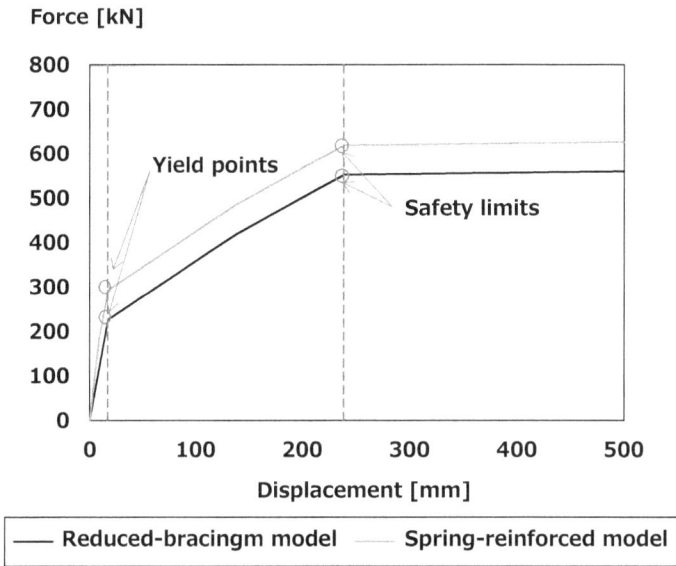

Figure 12.15 Chart indicating relation between force and displacement related to frame models: Bracing-reduced model and spring-reinforced model.

model and the spring-reinforced model, the latter being a modified version of the bracing-reduced model that was reinforced by the spring oil dampers. In terms of the models' design criteria, they were basically the same but for minor deviations between the modified models and the standard model; the same figure of 1/30 was uniformly applied as with the standard model. For the spring-reinforced model, I_s and q were calculated as explained below. For this calculation, the ultimate horizontal resistant force was computed based on the load increase analysis, and the same analysis conditions that were applied to the standard model were used, so they are not restated here. The spring-reinforced model's I_s was 0.62 while its q was 0.75. Hence, in terms of the assessment of this spring-reinforced model's seismic performance based on the computed values, there would be a significant risk that the structure might collapse or be destroyed in response to seismic vibration and shock (assuming $0.3 \leq I_s < 0.6$ and $0.5 \leq q < 1.0$).[1]

$$E_0 = \frac{Q_u F}{WA} = \frac{293.63 \times 3.3}{1560.34 \times 1.0} = 0.62$$

$$I_s = \frac{E_0}{F_{es} Z R_t} = \frac{0.62}{1.0 \times 1.0 \times 1.0} = 0.62$$

$$q = \frac{Q_u}{0.25 F_{es} WAZR_t} = \frac{293.63}{0.25 \times 1.0 \times 1560.34 \times 1.0 \times 1.0 \times 1.0} = 0.75$$

The converted I_s and q values were calculated by multiplying I_s and q by the reciprocal of the response reduction factor relating to the spring oil dampers' attenuation effect, respectively, applying the formulas (1) through (6). The results of this calculation are presented in Table 12.9, from which it could be discerned that the converted q value would be at or above 1.0 if the attenuation effect of the spring oil dampers was considered.

Then, time history response analysis was performed by applying different acceleration waveforms to the bracing-reduced model and the spring-reinforced model. As the analysis conditions applied were the same as those previously used on the standard model, they are not restated here. The results of the time history response analysis are shown in Table 12.10 as the models' maximum relative story drift angles specific to the input acceleration waveforms, and in Table 12.11 as the models' maximum absolute response acceleration. In addition, Table 12.12 provides a breakdown of the energy absorption rates, while Figure 12.16 presents a graph indicating the relation between story shear force and relative story drift angle specific to each model after the BCJ-L2 input. Table 12.10 shows that the maximum relative story drift angle of the bracing-reduced model was 1/38 rad while that of the spring-reinforced model was 1/105 rad. As the design criterion at the safety limit was set to 1/30 rad for each model, each model's maximum relative story drift angle apparently did not exceed the design criterion. However, as the bracing-reduced model's maximum relative story drift angle was fairly close to the design criterion, and the acceleration waveforms selected for this

Table 12.9 Computation of response reduction factor D_h

M [t]	159.11
k (pre-reinforcement) [kN/m]	2313
k (post-reinforcement) [kN/m]	2587
c_{od} [kN•s/m]	334
k_{od} [kN/m]	55000
c_{od}' [kN•s/m]	331.45
c_b [kN•s/m]	332
k_p [kN/m]	100000
c_{bod} [kN•s/m]	69.63
h_{bod}	0.054
h_0	0.02
heq	0.074
D_h	0.725
$1/D_h$	1.38
I_s	0.62
q	0.75
I_{sd}	0.86
q_d	1.04

Table 12.10 Maximum relative story drift angle data of different models [rad]

	Bracing-reduced model	Spring-reinforced model	Standard model
El Centro	1/54	1/119	1/51
Taft	1/40	1/105	1/61
Hachinohe	1/59	1/273	1/98
BCJ-L2	1/38	1/110	1/70

Table 12.11 Maximum absolute response acceleration data of different models [cm/s^2]

	Bracing-reduced model	Spring-reinforced model	Standard model
El Centro	265.93	312.76	398.29
Taft	306.58	323.58	375.33
Hachinohe	255.78	268.70	328.53
BCJ-L2	315.37	317.75	360.54

Table 12.12 Breakdown of anergy absorption rate data: BCJ-L2 [%]

	Analysis cases		
Type	Bracing-reduced model	Spring-reinforced model	Standard model
E_d/E_i	10.63	3.70	11.09
E_m/E_i	—	60.86	—
E_s/E_i	89.20	5.34	88.87
E_v/E_i	—	9.15	—
E_h/E_i	—	20.56	—

where: E_i = energy input; E_d = energy absorption due to structural attenuation; E_m = energy absorption according to the Maxwell model; E_s = energy absorption according to the spring model; E_v = energy absorption according to the Voigt model; and E_h = energy absorption according to the friction damper model.

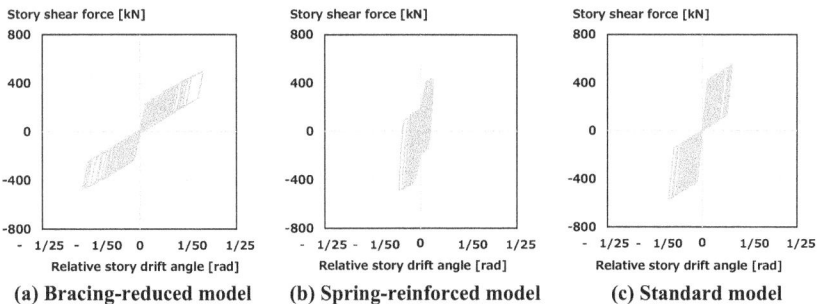

(a) Bracing-reduced model   (b) Spring-reinforced model   (c) Standard model

Figure 12.16 Chart indicating relation between story shear force and relative story drift angle: BCJ-L2.

study did not consider the amplification factor that might arise depending on the geotechnical properties of the ground where the structure would be built, the bracing-reduced model might pose the risk of collapsing or being destroyed in response to seismic vibration and shock, which was basically the same conclusion drawn during the assessment of the model's seismic performance based on I_s and q. Turning attention to the spring-reinforced model, its maximum relative story drift angle was 1/105 rad, which was close to the standard model's maximum relative story drift angle of 1/98. In terms of I_s and q comparison, the spring-reinforced model's converted I_s and q values – reflecting the response reduction factor – were 0.89 and 1.08, respectively, while the standard model's I_s was 0.90 and q was 1.09, so they were fairly close. However, as any frame model's response value following an observed waveform input sometimes varies significantly depending on the frame model's natural period, the BCJ-L2 waveform was applied to see if the outcome would be any different in terms of maximum relative story drift angle. This artificial waveform was selected for its known characteristic of being less affected by the frequency components, and for the tendency of its displacement response spectrum to increase almost linearly along with the natural period. The resulting maximum relative story drift angle was 1/110 rad for the spring-reinforced model and 1/70 rad for the standard model, so the spring-reinforced model yielded a significantly lower figure between the two. Shifting attention to how the models had fared in terms of maximum relative story drift angle following the acceleration waveform inputs derived from the observed seismic waves, the spring-reinforced model yielded lower figures than did the standard model overall, which could have been caused by I_s and q not corresponding to each other. However, as the oil dampers' bilinear characteristic was switched to an equivalent linear one in a simplified manner when the spring-reinforced model's converted I_s and q were calculated, the above phenomenon might have been due to an underestimation of the response reduction effect of the oil dampers' attenuation. It was then inferred from the above that the converted I_s and q values could fluctuate depending on how the oil dampers' attenuation characteristics had been assessed. Considering all this, the test method that was employed in this study might have been a simple and safe static assessment technique that considered the response reduction effect of the spring oil dampers' attenuation characteristics, erring on the side of caution. Examination of the maximum absolute response acceleration of each model presented in Table 12.11 revealed that the spring-reinforced model had a lower maximum absolute response acceleration than the standard model. In addition, as the bracing-reduced model exhibited a smaller story shear force across its frame model comprised of the bracing components in part, its maximum absolute response acceleration might have been of such a low figure. Further, Table 12.12 indicates that the spring-reinforced model had most of its energy input absorbed by the spring oil dampers.

Then when the graphs indicating the relation between story shear force and relative story drift angle shown in Figure 12.16 were analyzed, it was determined that the hysteretic characteristics of the bracing-reduced model and the standard model had exhibited such restoration force properties in their bracing components that were evidently representative of a bilinear-slip model. As for the spring-reinforced model, because it had had the spring oil dampers added to the bracing components, its energy absorption increased especially in the second and fourth quadrants.

12.5 SUMMARY

In this chapter, the method has been proposed that can be applied to statically assess a structure's seismic performance in response reduction through seismic attenuation involving spring oil dampers, based on the seismic index of structure (I_s) and ultimate horizontal resistant force (q) as specified in the seismic diagnosis.

First, Section 12.2 provided the standard building model's specifications and seismic performance data.

Then, Section 12.3 described the simple computation method that could be employed to calculate the converted I_s and q values that reflected the response reduction effect of the spring oil dampers through attenuation.

Finally, Section 12.4 proposed the modified versions of the standard building model that were introduced in Section 12.2, of which one had reduced bracing components while the other, too, had reduced bracing but also had oil dampers added for extra reinforcement. Then, the converted I_s and q value of each of these frame models were calculated to compare the results of their time history response analysis, pitting the two modified frame models against the standard frame model from Section 12.2. While the converted I_s and q values of these models – which reflected the response reduction effect of the spring oil dampers through attenuation – might have been underestimated in the models, each model's response values and its assessment based on the aforementioned performance indexes were roughly on par with each other. Hence, it was determined that the proposed method was an effective static assessment technique that could be used to evaluate the response reduction effect of the spring oil dampers through their attenuation properties based on the indexes as specified in the seismic diagnosis in a simplified and sufficiently safe manner.

REFERENCES

1. Japan Building Disaster Prevention Association (JBDPA): *Guidelines and Commentaries on the Seismic Diagnosis and Seismic Retrofitting of Existing Steel-Framed Structures for Compliance with the Act on Promotion of Seismic Retrofitting of Buildings*, 2011 revision. (in Japanese)

2. *SNAP Ver.7 Technical Manual.*
3. S Otani: *A Computer Program for Inelastic Analysis of R/C Frames to Earthquakes: A Report on a Research Project Sponsored by the National Science Foundation, Research Grant GI-29934*, University of Illinois at Urbana-Champaign, Nov. 1974.
4. Akenori Shibata: *Saishin Taishin-kōzō Kaiseki* (Dynamic Analysis of Earthquake Resistant Structures), 3rd Edition, Morikita Publishing Co., Ltd, Dec. 2014.
5. The Japan Society of Seismic Isolation (JSSI): *Passive Seismic Control Structure Design and Construction Manual*, 2nd Edition, July 2007. (in Japanese)

Chapter 13

Conclusion to Part 2

This research project has accomplished the following two main goals. First, it proposed the spring oil damper configuration comprised of shear-resistant spring units – each consisting of a low-loss viscoelastic material with very limited temperature-induced fluctuation in shear-resisting stiffness performance and a steel plate – and friction mechanisms, joined serially, with an oil damper set up in parallel. Then, the proposed spring oil damper configuration was put to the test involving the models of steel-framed low-stiffness buildings that were designed based on the now-obsolete seismic standards that would be used as factories or warehouses, to see how the proposed configuration would fare in comparison to the conventional reinforcement technique that would install braces alongside oil dampers, which demonstrated that the proposed use of spring oil dampers could not only reduce the number of reinforcement locations required on a given structure but also help save the cost of reinforcement components.

The following is a summary of the new knowledge and insights obtained during the project as described in each of the chapters.

Chapter 9 explained the spring oil dampers' configuration and presented the results of their standalone performance tests, from which their basic characteristics, multi-factor dependence, and durability after having been put through a series of tests were determined. At one point, the spring oil dampers were tested with oil drained from the oil dampers within to test the independent performance of the built-in spring units, each of which was comprised of a viscoelastic spring and a friction mechanism. After applying sine waves of different cycles in the ensuing test, it was determined that the spring units had performance capability basically meeting the design-specified values and that their performance would not fluctuate much in response to change in excitation amplitude or velocity. Then, the durability test was conducted that simulated repetitive seismic wave inputs that would typically occur during major earthquakes as well as repetitive vibration inputs mimicking normal winds and minor to medium-intensity earthquakes, which demonstrated that the spring units were sufficiently durable. The random wave vibration test followed, which involved applying earthquake-like random seismic waves to the spring units, and it was observed

DOI: 10.1201/9781003290261-15

if they had operated smoothly. Then, the performance test was conducted on the spring oil dampers, for which their performance had been adjusted to suit the test apparatus specifications, and the performance of the spring oil dampers as whole devices was verified. It was then determined that the viscoelastic material – which was incorporated into each spring oil damper to provide stiffness support – had absorbed through attenuation roughly 10 to 30% of all energy absorbed by the entire spring oil damper during such vibrations that would be caused by normal winds as well as by minor to medium-intensity earthquakes.

Chapter 10 introduced the two types of analysis models that could be applied to spring oil dampers, the differentiating factor being how the viscoelastic springs were modeled. Then, the test results and the analysis results calculated using the different computation models were compared to determine the validity of the analysis model that was proposed. This improved model – which reflected the viscoelastic springs' attenuation properties – was determined to have tracked the actual test results more closely than did the simple computation model, in terms of energy absorption breakdown and deviation rates, across all test conditions, while achieving an accuracy of around ±10%, partly because the simple computation model did not consider the viscoelastic springs' attenuation properties. Hence, it was demonstrated from the comparison of the test and analysis results that the improved model could better predict the energy absorption by the viscoelastic springs and more accurately simulate the test results, as the improved model had the viscoelastic springs inside the spring oil dampers modeled according to the Voigt model instead of the linear spring model.

Chapter 11 described the time history response analysis that was performed on the single degree-of-freedom model of an actual factory building, which determined how the different analysis models of spring oil dampers would affect the response values that arose from the single degree-of-freedom model, whereby the efficacy of the spring oil dampers was verified. Then Lv. 2 acceleration waveform inputs were given to the building model, applying the simple computation model and the improved model one after the other, which adopted the different methods of modeling the viscoelastic springs within the spring oil dampers, and it was determined that the response values between the two computation models did not show any significant difference. In addition, as it was assumed that each type of the reinforcement components would be installed in six different locations in the building model, the total number of reinforcement locations would be 12 for ODS and six for BODS and BODV, which demonstrated that the use of spring oil dampers this way could reduce the number of reinforcement locations while improving the structure's post-reinforcement performance. It was also determined that the use of spring oil dampers as proposed would be superior to the conventional reinforcement method of installing braces alongside oil dampers, as the proposed method could not only reduce the number of reinforcement locations but also save the cost of reinforcement components.

Chapter 12 proposed the method that could be applied to statically assess a structure's seismic performance in response reduction through seismic attenuation involving spring oil dampers, based on the seismic index of structure (I_s) and ultimate horizontal resistant force (q) as specified in the seismic diagnosis. While the converted I_s and q values of the different models – which reflected the response reduction effect of the spring oil dampers through attenuation – might have been underestimated in the models, each model's response values and its assessment based on the aforementioned performance indexes roughly corresponded to each other, thereby demonstrating that the proposed method was an effective static assessment technique that could be used to evaluate the response reduction effect of the spring oil dampers through their attenuation properties as per the indexes specified in the seismic diagnosis in a manner that was uncomplicated and erred on the side of safety.

Appendix 1 Example of implementation in a high-rise building (office building, Atago 2-chome Project (tentative name))

A1.1 BUILDING OVERVIEW

Building name:	Atago Green Hills Office Building
Location:	Minato-ku, Tokyo
Use:	Stores, offices
Design:	Mori Building Co., Ltd. First-class Architects Structural Planning Research Institute Co., Ltd.
Building area:	2,426.41 m²
Floor area:	86,844.05 m²
Number of floors:	3 basement floors, 41 floors above ground, 2 floors of towers
Height:	186.76 m
Structure type:	Steel structure
Basic land industry:	Direct foundation (solid foundation)
Vibration damping member:	Brace-type oil damper 500 kN type, 1000 kN type, total 688 units

Figure A1.1   Building exterior (left: office building/right: residential building).

A1.2 FEATURES OF THE BUILDING PLAN

Atago Green Hills is located around Seishoji Temple in Atago 2-chome, and two skyscrapers are being constructed. The building located on the south side is an office building with a height of about 190 m, and the building on the north side is a residential building with a height of about 160 m. Both buildings have a damping structure with a brace-type oil damper placed in the center core.

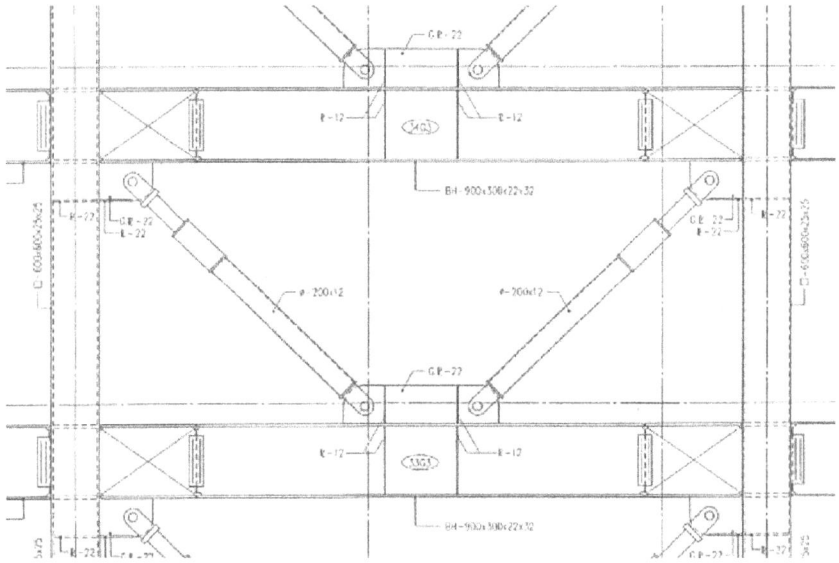

Figure A1.2 Brace-type oil damper installation.

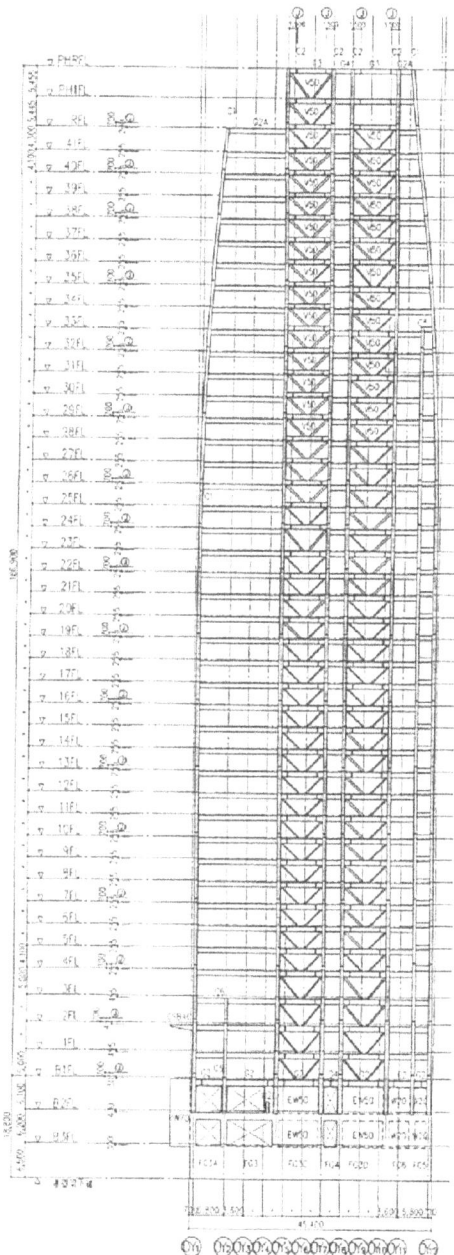

Figure A1.3 Maximum response interlayer displacement (level 2 excitation in X direction).

A1.3 DESIGN GOALS

- Rare seismic motion level
- Extremely rare seismic motion level

Table A1.1 Design goals

	Level 1[*1]	Level 2[*2]
Interlayer deformation angle	1/200 or less	1/100 or less
Plasticity rate	Within elastic range	2.0 or less

[*1] The level of earthquake occurring once per decades years in Japan

[*2] The level of earthquake occurring once per hundreds years in Japan

A1.4 SEISMIC RESPONSE ANALYSIS AND ANALYSIS RESULTS

In both the X and Y directions, after separating each layer into bending deformation (rocking) and shear deformation (sway) from the load increment analysis results, beam replacement was performed so that each layer had equivalent rigidity, and each layer was set to the trilinear restoring force characteristic. Then, time history response analysis (elasto-plasticity) was performed. The oil damper was evaluated by the Maxwell model and installed between the mass points. Tables A1.2 and A1.3 show the analysis results for the TAFT waves (Level 1, Level 2) with the strictest response values.

Table A1.2. Analysis results (level 1)

	X direction	Y direction
Max. interlayer deformation angle	1/495 (7F)	1/419 (19F)
Max. layer shear force coefficient (B1F)	0.037	0.0389
Ratio to design shear force (max. value)	0.78 (37F)	0.82 (13F)

Table A1.3 Analysis results (level 2)

	X direction	Y direction
Max. interlayer deformation angle	1/249 (7F)	1/207 (19F)
Max. layer shear force coefficient (B1F)	0.0737	0.0777
Max. layer plasticity rate	0.629 (3F)	0.707 (3F)

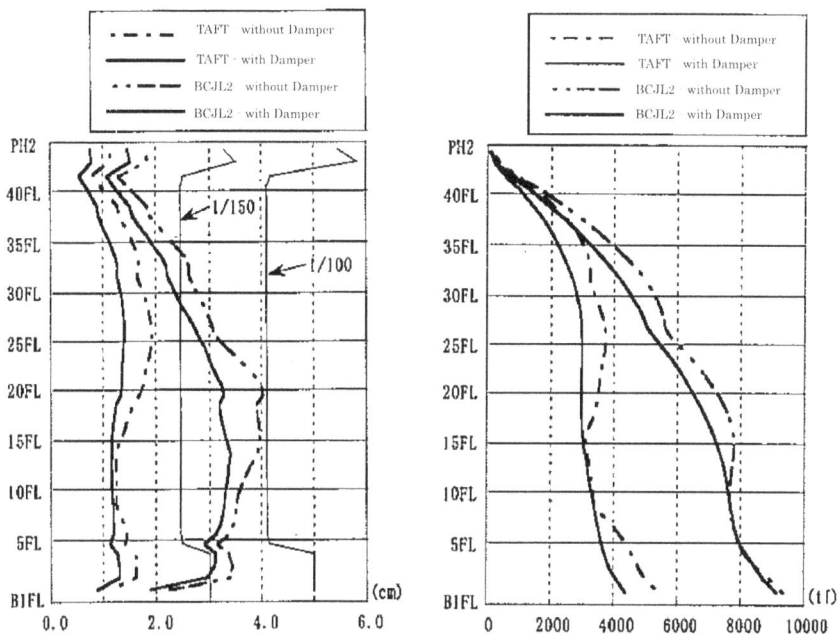

Figure A1.4 Maximum response interlayer shear (level 2 excitation in X direction).

Appendix 2 Example of implementation in a reinforced building (Aizu-Tajima Joint Government Building, Fukushima Prefecture)

A2.1 BUILDING OVERVIEW

Building name:	Fukushima Prefecture Aizu-Tajima Joint Government Building
location:	Fukushima Prefecture
Use:	Government building
Design:	Kozo Keikaku Kenkyusho Co., Ltd. (seismic retrofitting)
Building area:	858.17 m²
Floor area:	3,383.06 m²
Number of floors:	4 floors above ground, 2 floors of tower
Height:	19.6 m
Structure type:	Reinforced concrete structure
Basic land industry:	Pile foundation
Damper:	Brace-type oil damper
	48 units of 500 kN type

Figure A2.1 Building exterior.

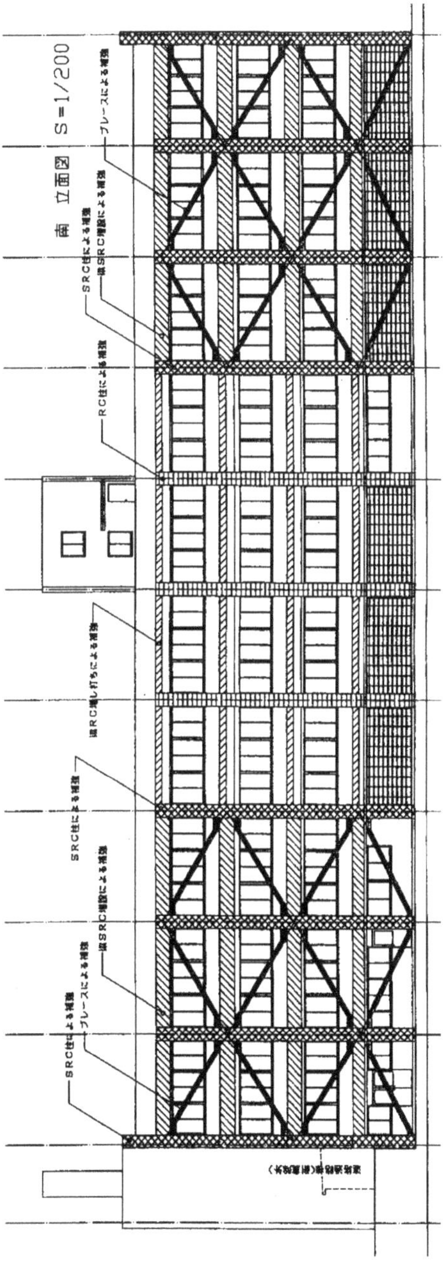

Figure A2.2 Elevation (after renovation).

A2.2 DESIGN CONCEPT

[Realization of construction while staying]

Reinforcing members were added mainly from the outside of the building to realize the construction while being able to reduce the burden of moving or relocating inside the facility. Specifically, the column-beam cross section in the girder direction was increased to improve the bearing capacity and toughness of the members, and at the same time, the bearing capacity was also improved in the beam-to-beam direction.

[Reduction of seismic response by damping damper]

A brace-type oil damper was attached externally to reduce the seismic response. The design goals at this time were to set the interlayer deformation angle during a large earthquake to 1/200 or less and the layer plasticity rate to 2.0 or less.

[Reinforcement of foundation]

To deal with the foundation reaction force, during an earthquake in the direction between the beams and the increase in building weight due to reinforcement, the foundation was reinforced by adding piles. At this time, thorough consideration was given to realizing the construction while staying, such as selecting a ready-made pile construction method that can be constructed with small-scale heavy machinery.

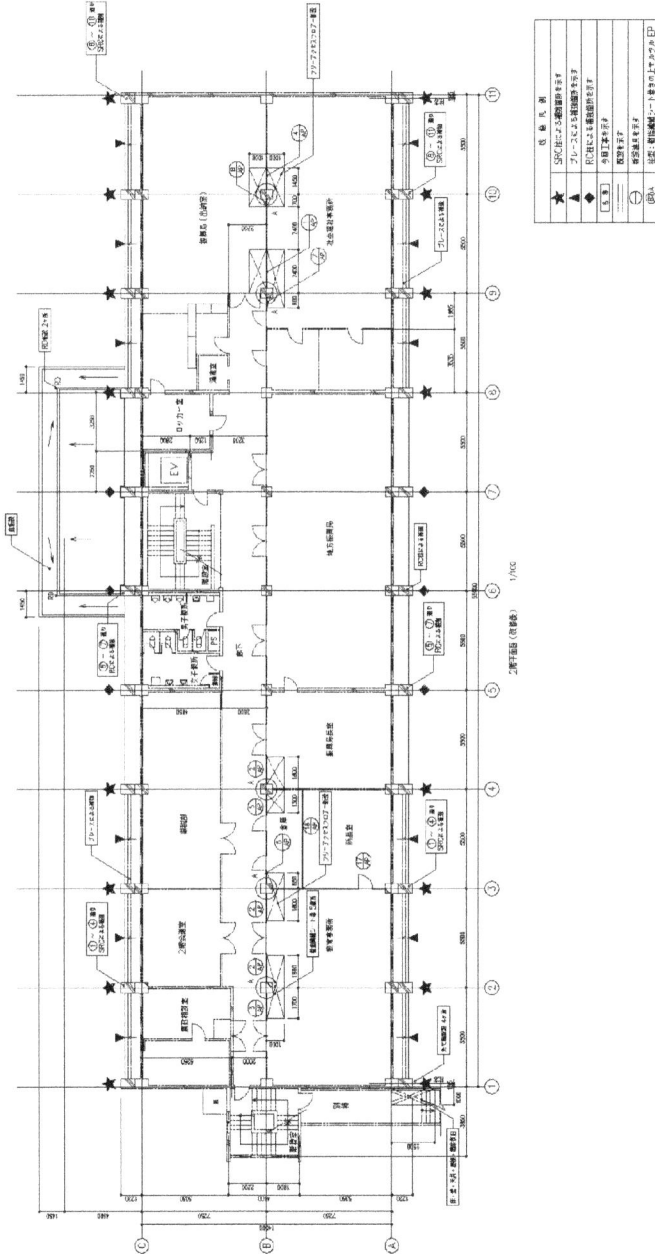

Figure A2.3 2nd floor plan (after renovation).

A2.3 SEISMIC PERFORMANCE TARGET

• Extremely rare seismic motion level

Table A2.1 Design goals

	Level 2[*1]
Interlayer deformation angle	1/200 or less
Plasticity rate	2.0 or less

[*1] The level of earthquake occurring once per hundreds years in Japan

A2.4 SEISMIC RESPONSE ANALYSIS AND ANALYSIS RESULTS

In both the X and Y directions, the layer shear force - interlayer displacement curve of each floor obtained by the three-dimensional static elasto-plastic analysis is modeled in a trilinear manner so that the area is equivalent by the equivalent shear model, and the time history response analysis (elasto-plasticity) is performed. The oil damper was evaluated by a nonlinear Maxwell model and installed between mass points. Table A2.2 shows the analysis results for the X direction with damping reinforcement.

Table A2.2 Analysis results

	X direction	
	Existing buildings	After damping reinforcement
Interlayer deformation angle (interlayer displacement)	1/72 (3F) (4.87cm) BCJL2	1/272 (3F) (1.26cm) TAFT
Ratio to design shear force (max. value)	1.3 (1F) TAFT	1.65 (1F) TAFT
Interlayer speed (max. value)	–	14.8kine (3F) EL CENTRO
During enveloping shear force Max. part plasticity rate (load incremental analysis)	–	1.35 (Wall, 1F)

Appendix 3 The low-loss viscoelastic material's dependence on temperature and vibration frequency

Appended Figure A3.1 indicates the low-loss viscoelastic material's storage modulus of elasticity, loss modulus of elasticity, complex modulus of elasticity, and loss factor by temperature, while Appended Figure A3.2 presents its same properties by vibration frequency.

From Appended Figure A3.1, it can be discerned that the material's storage modulus of elasticity and loss factor fluctuated only slightly in the temperature range of 10° to 30°, which would be its scope of application in terms of the dynamic excitation conditions of vibration control components. In particular, the storage modulus of elasticity – which would significantly affect the stiffness of vibration control components – hardly fluctuated in the aforementioned range. In addition, according to Appended Figure A3.2, each of the material's properties only fluctuated slightly in the 0.25 Hz to 10 Hz range, which is considered to be the typical range of structures' natural frequencies.

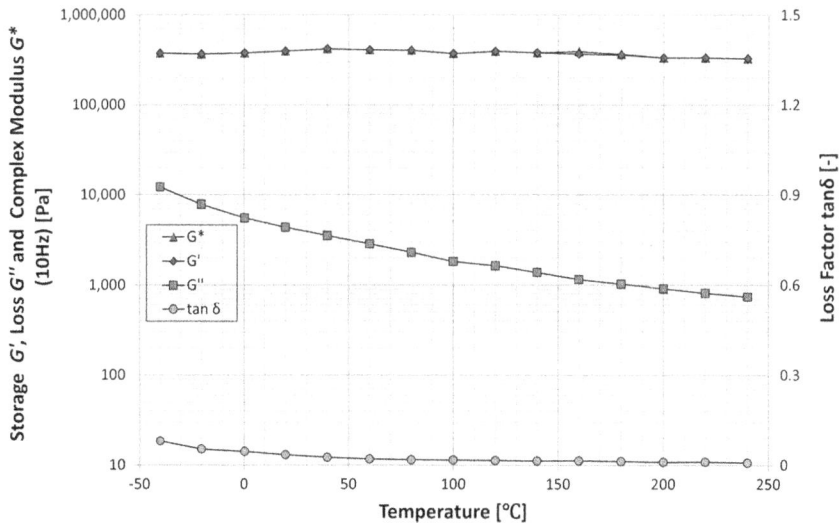

Figure A3.1 Low-loss viscoelastic material's properties by temperature.

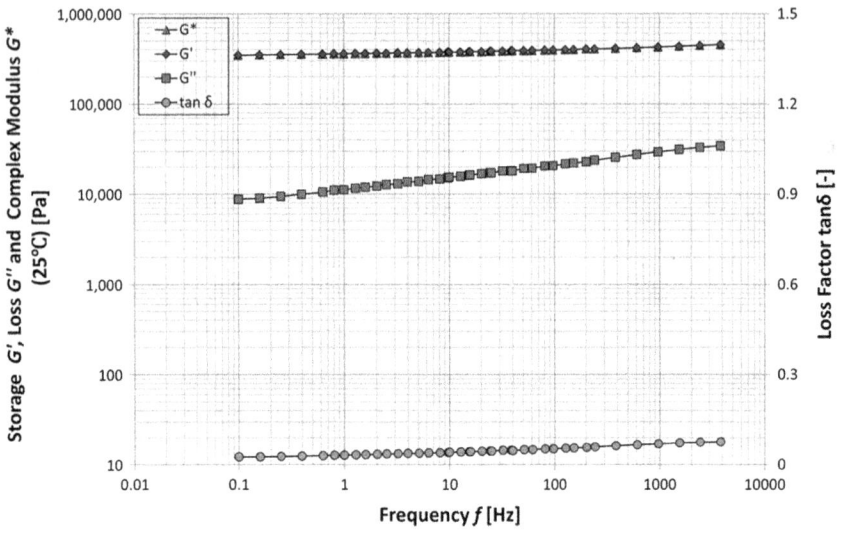

Figure A3.2 Low-loss viscoelastic material's properties by frequency.

Index

For Product Safety Concerns and Information please contact our EU
representative GPSR@taylorandfrancis.com
Taylor & Francis Verlag GmbH, Kaufingerstraße 24, 80331 München, Germany